The Proton

" The Partner of The Neutron "

Edited by Paul F. Kisak

Contents

1 Proton   **1**

   1.1 Description   1

   1.2 History   2

   1.3 Stability   2

   1.4 Quarks and the mass of the proton   3

   1.5 Charge radius   4

   1.6 Interaction of free protons with ordinary matter   4

   1.7 Proton in chemistry   4

       1.7.1 Atomic number   4

       1.7.2 Hydrogen ion   5

       1.7.3 Proton nuclear magnetic resonance (NMR)   5

   1.8 Human exposure   5

   1.9 Antiproton   6

   1.10 See also   6

   1.11 References   6

   1.12 External links   8

2 Subatomic particle   **11**

   2.1 Classification   11

       2.1.1 By statistics   11

       2.1.2 By composition   11

       2.1.3 By mass   12

   2.2 Other properties   13

   2.3 Dividing an atom   13

   2.4 History   13

   2.5 See also   13

   2.6 References   14

   2.7 Further reading   14

   2.8 External links   15

3 Elementary particle   **16**

  3.1 Overview 17

  3.2 Common elementary particles 18

  3.3 Standard Model 19

      3.3.1 Fundamental fermions 19

      3.3.2 Fundamental bosons 21

  3.4 Beyond the Standard Model 22

      3.4.1 Grand unification 22

      3.4.2 Supersymmetry 22

      3.4.3 String theory 22

      3.4.4 Technicolor 23

      3.4.5 Preon theory 23

      3.4.6 Acceleron theory 23

  3.5 See also 23

  3.6 Notes 23

  3.7 Further reading 24

      3.7.1 General readers 24

      3.7.2 Textbooks 25

  3.8 External links 25

4 List of particles   **26**

  4.1 Elementary particles 26

      4.1.1 Fermions 26

      4.1.2 Bosons 27

      4.1.3 Hypothetical particles 27

  4.2 Composite particles 28

      4.2.1 Hadrons 28

      4.2.2 Atomic nuclei 29

      4.2.3 Atoms 30

      4.2.4 Molecules 31

  4.3 Condensed matter 31

  4.4 Other 32

  4.5 Classification by speed 33

  4.6 See also 33

  4.7 References 34

5 Standard Model   **35**

  5.1 Historical background 36

  5.2 Overview 36

5.3 Particle content 36

    5.3.1 Fermions 37

    5.3.2 Gauge bosons 38

    5.3.3 Higgs boson 38

    5.3.4 Total particle count 40

5.4 Theoretical aspects 40

    5.4.1 Construction of the Standard Model Lagrangian 40

5.5 Fundamental forces 41

5.6 Tests and predictions 42

5.7 Challenges 42

5.8 See also 43

5.9 Notes and references 43

5.10 References 44

5.11 Further reading 45

5.12 External links 46

6 Charge radius   **48**

6.1 Definition 48

6.2 History 49

6.3 Modern measurements 49

6.4 References 49

7 Proton decay   **51**

7.1 Baryogenesis 51

7.2 Experimental evidence 52

7.3 Theoretical motivation 53

7.4 Decay operators 53

    7.4.1 Dimension-6 proton decay operators 53

    7.4.2 Dimension-5 proton decay operators 53

    7.4.3 Dimension-4 proton decay operators 54

7.5 See also 54

7.6 References 55

7.7 Further reading 55

7.8 External links 55

8 Radioactive decay   **56**

8.1 History of discovery 57

8.2 Early health dangers 58

    8.2.1 X-rays 58

8.2.2 Radioactive substances 59

8.2.3 Radiation protection 60

8.3 Units of radioactivity 60

8.4 Types of decay 61

8.5 Radioactive decay rates 64

8.6 Mathematics of radioactive decay 65

8.6.1 Universal law of radioactive decay 65

8.6.2 Corollaries of the decay laws 68

8.6.3 Decay timing: definitions and relations 69

8.6.4 Example 70

8.7 Changing decay rates 70

8.8 Theoretical basis of decay phenomena 71

8.9 Occurrence and applications 71

8.10 Origins of radioactive nuclides 72

8.11 Decay chains and multiple modes 72

8.12 Associated hazard warning signs 73

8.13 See also 73

8.14 Notes 74

8.15 References 74

8.15.1 Inline 74

8.15.2 General 76

8.16 External links 76

9 Electron capture **81**

9.1 History 83

9.2 Reaction details 83

9.3 Common examples 83

9.4 References 83

9.5 External links 84

10 Quantum chromodynamics **85**

10.1 Terminology 85

10.2 History 86

10.3 Theory 87

10.3.1 Some definitions 87

10.3.2 Additional remarks: duality 87

10.3.3 Symmetry groups 88

10.3.4 Lagrangian 88

10.3.5 Fields 89

10.3.6 Dynamics 90

10.3.7 Area law and confinement 90

10.4 Methods 90

10.4.1 Perturbative QCD 90

10.4.2 Lattice QCD 90

10.4.3 1/N expansion 90

10.4.4 Effective theories 91

10.4.5 QCD sum rules 91

10.4.6 Nambu–Jona-Lasinio model 91

10.5 Experimental tests 91

10.6 Cross-relations to solid state physics 92

10.7 See also 93

10.8 References 93

10.9 Further reading 94

10.10External links 94

11 Proton therapy **96**

11.1 Description 97

11.2 History 98

11.3 Application 98

11.4 Comparison with other treatments 99

11.4.1 X-ray radiotherapy 99

11.4.2 Surgery 101

11.5 Side effects and risks 101

11.6 Costs 101

11.7 Treatment centers 101

11.7.1 United States 102

11.7.2 Outside the USA 102

11.8 United Kingdom 102

11.9 See also 103

11.10References 103

11.11Further reading 106

11.12External links 107

12 Hydron (chemistry) **108**

12.1 Isotopes of hydron 108

12.2 History of the term 109

12.3 See also 109

12.4 References 109

13 Proton nuclear magnetic resonance   **111**

    13.1   Chemical shifts 112

    13.2   Signal strength 112

    13.3   Spin-spin couplings 113

    13.4   Carbon satellites and spinning sidebands 115

    13.5   See also 116

    13.6   References 116

    13.7   External links 116

14 Proton–proton chain reaction   **117**

    14.1   History of the theory 117

    14.2   The pp chain reaction 117

        14.2.1   The pp I branch 118

        14.2.2   The pp II branch 118

        14.2.3   The pp III branch 119

        14.2.4   The pp IV (Hep) branch 119

        14.2.5   Energy release 119

    14.3   The pep reaction 119

    14.4   See also 119

    14.5   References 120

15 Proton spin crisis   **125**

    15.1   Background 125

    15.2   The experiment 125

    15.3   Recent work 125

    15.4   References 126

    15.5   External links 126

16 Antiproton   **127**

    16.1   Occurrence in nature 127

        16.1.1   List of recent antiproton cosmic ray detection experiments 128

    16.2   Modern experiments and applications 128

    16.3   See also 129

    16.4   References 129

    16.5   Text and image sources, contributors, and licenses 130

        16.5.1   Text 130

        16.5.2   Images 135

        16.5.3   Content license 138

Chapter 1

Proton

This article is about the proton as a subatomic particle. For other uses, see Proton (disambiguation).

The **proton** is an elementary subatomic particle, symbol p or p+, with a positive electric charge of +1e elementary charge and mass slightly less than that of a neutron. Protons and neutrons, each with mass approximately one atomic mass unit, are collectively referred to as "nucleons". One or more protons are present in the nucleus of an atom. The number of protons in the nucleus is referred to as its atomic number. Since each element has a unique number of protons, each element has its own unique atomic number. The word *proton* is Greek for "first", and this name was given to the hydrogen nucleus by Ernest Rutherford in 1920. In previous years Rutherford had discovered that the hydrogen nucleus (known to be the lightest nucleus) could be extracted from the nuclei of nitrogen by collision. The proton was therefore a candidate to be a fundamental particle and a building block of nitrogen and all other heavier atomic nuclei.

In the modern Standard Model of particle physics, the proton is a hadron, and like the neutron, the other nucleon (particle present in atomic nuclei), is composed of three quarks. Although the proton was originally considered a fundamental particle, it is composed of three valence quarks: two up quarks and one down quark. The rest masses of the quarks contribute only about 1% of the proton's mass, however.[2] The remainder of the proton mass is due to the kinetic energy of the quarks and to the energy of the gluon fields that bind the quarks together. Because the proton is not a fundamental particle, it possesses a physical size; the radius of the proton is about 0.84–0.87 fm.[3]

At sufficiently low temperatures, free protons will bind to electrons. However, the character of such bound protons does not change, and they remain protons. A fast proton moving through matter will slow by interactions with electrons and nuclei, until it is captured by the electron cloud of an atom. The result is a protonated atom, which is a chemical compound of hydrogen. In vacuum, when free electrons are present, a sufficiently slow proton may pick up a single free electron, becoming a neutral hydrogen atom, which is chemically a free radical. Such "free hydrogen atoms" tend to react chemically with many other types of atoms at sufficiently low energies. When free hydrogen atoms react with each other, they form neutral hydrogen molecules (H_2), which are the most common molecular component of molecular clouds in interstellar space. Such molecules of hydrogen on Earth may then serve (among many other uses) as a convenient source of protons for accelerators (as used in proton therapy) and other hadron particle physics experiments that require protons to accelerate, with the most powerful and noted example being the Large Hadron Collider.

1.1 Description

Protons are spin-½ fermions and are composed of three valence quarks,[4] making them baryons (a sub-type of hadrons). The two up quarks and one down quark of the proton are held together by the strong force, mediated by gluons.[5]:21–22 A modern perspective has the proton composed of the valence quarks (up, up, down), the gluons, and transitory pairs of sea quarks. The proton has an approximately exponentially decaying positive charge distribution with a mean square radius of about 0.8 fm.[6]

Protons and neutrons are both nucleons, which may be bound together by the nuclear force to form atomic nuclei. The nucleus of the most common isotope of the hydrogen atom (with the chemical symbol "H") is a lone proton. The nuclei of the heavy hydrogen isotopes deuterium and tritium contain one proton bound to one and two neutrons, respectively. All other types of atomic nuclei are composed of two or more protons and various numbers of neutrons.

1.2   History

The concept of a hydrogen-like particle as a constituent of other atoms was developed over a long period. As early as 1815, William Prout proposed that all atoms are composed of hydrogen atoms (which he called "protyles"), based on a simplistic interpretation of early values of atomic weights (see Prout's hypothesis), which was disproved when more accurate values were measured.[7]:39–42

In 1886, Eugen Goldstein discovered canal rays (also known as anode rays) and showed that they were positively charged particles (ions) produced from gases. However, since particles from different gases had different values of charge-to-mass ratio (e/m), they could not be identified with a single particle, unlike the negative electrons discovered by J. J. Thomson.

Following the discovery of the atomic nucleus by Ernest Rutherford in 1911, Antonius van den Broek proposed that the place of each element in the periodic table (its atomic number) is equal to its nuclear charge. This was confirmed experimentally by Henry Moseley in 1913 using X-ray spectra.

In 1917, (in experiments reported in 1919) Rutherford proved that the hydrogen nucleus is present in other nuclei, a result usually described as the discovery of the proton.[8] Rutherford had earlier learned to produce hydrogen nuclei as a type of radiation produced as a product of the impact of alpha particles on nitrogen gas, and recognize them by their unique penetration signature in air and their appearance in scintillation detectors. These experiments were begun when Rutherford had noticed that, when alpha particles were shot into air (mostly nitrogen), his scintillation detectors showed the signatures of typical hydrogen nuclei as a product. After experimentation Rutherford traced the reaction to the nitrogen in air, and found that when alphas were produced into pure nitrogen gas, the effect was larger. Rutherford determined that this hydrogen could have come only from the nitrogen, and therefore nitrogen must contain hydrogen nuclei. One hydrogen nucleus was being knocked off by the impact of the alpha particle, producing oxygen-17 in the process. This was the first reported nuclear reaction, $^{14}N + \alpha \rightarrow {}^{17}O + p$. (This reaction would later be observed happening directly in a cloud chamber in 1925).

Rutherford knew hydrogen to be the simplest and lightest element and was influenced by Prout's hypothesis that hydrogen was the building block of all elements. Discovery that the hydrogen nucleus is present in all other nuclei as an elementary particle, led Rutherford to give the hydrogen nucleus a special name as a particle, since he suspected that hydrogen, the lightest element, contained only one of these particles. He named this new fundamental building block of the nucleus the *proton,* after the neuter singular of the Greek word for "first", πρῶτον. However, Rutherford also had in mind the word *protyle* as used by Prout. Rutherford spoke at the British Association for the Advancement of Science at its Cardiff meeting beginning 24 August 1920.[9] Rutherford was asked by Oliver Lodge for a new name for the positive hydrogen nucleus to avoid confusion with the neutral hydrogen atom. He initially suggested both *proton* and *prouton* (after Prout).[10] Rutherford later reported that the meeting had accepted his suggestion that the hydrogen nucleus be named the "proton", following Prout's word "protyle".[11] The first use of the word "proton" in the scientific literature appeared in 1920.[12]

1.3   Stability

Main article: Proton decay

The free proton (a proton not bound to nucleons or electrons) is a stable particle that has not been observed to break down spontaneously to other particles. Free protons are found naturally in a number of situations in which energies or temperatures are high enough to separate them from electrons, for which they have some affinity. Free protons exist in plasmas in which temperatures are too high to allow them to combine with electrons. Free protons of high energy and velocity make up 90% of cosmic rays, which propagate in vacuum for interstellar distances. Free protons are emitted directly from atomic nuclei in some rare types of radioactive decay. Protons also result (along with electrons and antineutrinos) from the radioactive decay of free neutrons, which are unstable.

The spontaneous decay of free protons has never been observed, and the proton is therefore considered a stable particle. However, some grand unified theories of particle physics predict that proton decay should take place with lifetimes of the order of 10^{36} years, and experimental searches have established lower bounds on the mean lifetime of the proton for various assumed decay products.[13][14][15]

Experiments at the Super-Kamiokande detector in Japan gave lower limits for proton mean lifetime of 6.6×10^{33} years for decay to an antimuon and a neutral pion, and 8.2×10^{33} years for decay to a positron and a neutral pion.[16] Another experiment at the Sudbury Neutrino Observatory in Canada searched for gamma rays resulting from residual nuclei resulting from the decay of a proton from oxygen-16. This experiment was designed to detect decay to any product, and established a lower limit to the proton lifetime of 2.1×10^{29} years.[17]

However, protons are known to transform into neutrons through the process of electron capture (also called inverse beta decay). For free protons, this process does not occur spontaneously but only when energy is supplied. The equation is:

$$\text{p+ + e-} \rightarrow \text{n} + \nu_e$$

The process is reversible; neutrons can convert back to protons through beta decay, a common form of radioactive decay. In fact, a free neutron decays this way, with a mean lifetime of about 15 minutes.

1.4 Quarks and the mass of the proton

In quantum chromodynamics, the modern theory of the nuclear force, most of the mass of the proton and the neutron is explained by special relativity. The mass of the proton is about 80–100 times greater than the sum of the rest masses of the quarks that make it up, while the gluons have zero rest mass. The extra energy of the quarks and gluons in a region within a proton, as compared to the rest energy of the quarks alone in the QCD vacuum, accounts for almost 99% of the mass. The rest mass of the proton is, thus, the invariant mass of the system of moving quarks and gluons that make up the particle, and, in such systems, even the energy of massless particles is still measured as part of the rest mass of the system.

Two terms are used in referring to the mass of the quarks that make up protons: *current quark mass* refers to the mass of a quark by itself, while *constituent quark mass* refers to the current quark mass plus the mass of the gluon particle field surrounding the quark.[18]:285–286 [19]:150–151 These masses typically have very different values. As noted, most of a proton's mass comes from the gluons that bind the current quarks together, rather than from the quarks themselves. While gluons are inherently massless, they possess energy—to be more specific, quantum chromodynamics binding energy (QCBE)— and it is this that contributes so greatly to the overall mass of the proton (see mass in special relativity). A proton has a mass of approximately 938 MeV/c^2, of which the rest mass of its three valence quarks contributes only about 9.4 MeV/c^2; much of the remainder can be attributed to the gluons' QCBE.[20][21][22]

The internal dynamics of the proton are complicated, because they are determined by the quarks' exchanging gluons, and interacting with various vacuum condensates. Lattice QCD provides a way of calculating the mass of the proton directly from the theory to any accuracy, in principle. The most recent calculations[23][24] claim that the mass is determined to better than 4% accuracy, even to 1% accuracy (see Figure S5 in Dürr *et al.*[24]). These claims are still controversial, because the calculations cannot yet be done with quarks as light as they are in the real world. This means that the predictions are found by a process of extrapolation, which can introduce systematic errors.[25] It is hard to tell whether these errors are controlled properly, because the quantities that are compared to experiment are the masses of the hadrons, which are known in advance.

These recent calculations are performed by massive supercomputers, and, as noted by Boffi and Pasquini: "a detailed description of the nucleon structure is still missing because ... long-distance behavior requires a nonperturbative and/or numerical treatment..."[26] More conceptual approaches to the structure of the proton are: the topological soliton approach originally due to Tony Skyrme and the more accurate AdS/QCD approach that extends it to include a string theory of gluons,[27] various QCD-inspired models like the bag model and the constituent quark model, which were popular in the 1980s, and the SVZ sum rules, which allow for rough approximate mass calculations.[28] These methods do not have the same accuracy as the more brute-force lattice QCD methods, at least not yet.

1.5   Charge radius

Main article: Charge radius

The internationally accepted value of the proton's charge radius is 0.8768 fm (see orders of magnitude for comparison to other sizes). This value is based on measurements involving a proton and an electron.

However, since 5 July 2010, an international research team has been able to make measurements involving an exotic atom made of a proton and a negatively charged muon. After a long and careful analysis of those measurements, the team concluded that the root-mean-square charge radius of a proton is "0.84184(67) fm, which differs by 5.0 standard deviations from the CODATA value of 0.8768(69) fm".[29] In January 2013, an updated value for the charge radius of a proton—0.84087(39) fm—was published. The precision was improved by 1.7 times, but the difference with CODATA value persisted at 7σ significance.[30]

The international research team that obtained this result at the Paul Scherrer Institut (PSI) in Villigen (Switzerland) includes scientists from the Max Planck Institute of Quantum Optics (MPQ) in Garching, the Ludwig-Maximilians-Universität (LMU) Munich and the Institut für Strahlwerkzeuge (IFWS) of the Universität Stuttgart (both from Germany), and the University of Coimbra, Portugal.[31][32] They are now attempting to explain the discrepancy, and re-examining the results of both previous high-precision measurements and complicated calculations. If no errors are found in the measurements or calculations, it could be necessary to re-examine the world's most precise and best-tested fundamental theory: quantum electrodynamics.[31] The proton radius remains a puzzle as of early 2015.[33]

1.6   Interaction of free protons with ordinary matter

Main article: Proton therapy

Although protons have affinity for oppositely charged electrons, free protons must lose sufficient velocity (and kinetic energy) in order to become closely associated and bound to electrons, since this is a relatively low-energy interaction. High energy protons, in traversing ordinary matter, lose energy by collisions with atomic nuclei, and by ionization of atoms (removing electrons) until they are slowed sufficiently to be captured by the electron cloud in a normal atom.

However, in such an association with an electron, the character of the bound proton is not changed, and it remains a proton. The attraction of low-energy free protons to any electrons present in normal matter (such as the electrons in normal atoms) causes free protons to stop and to form a new chemical bond with an atom. Such a bond happens at any sufficiently "cold" temperature (i.e., comparable to temperatures at the surface of the Sun) and with any type of atom. Thus, in interaction with any type of normal (non-plasma) matter, low-velocity free protons are attracted to electrons in any atom or molecule with which they come in contact, causing the proton and molecule to combine. Such molecules are then said to be "protonated", and chemically they often, as a result, become so-called Bronsted acids.

1.7   Proton in chemistry

1.7.1   Atomic number

In chemistry, the number of protons in the nucleus of an atom is known as the atomic number, which determines the chemical element to which the atom belongs. For example, the atomic number of chlorine is 17; this means that each chlorine atom has 17 protons and that all atoms with 17 protons are chlorine atoms. The chemical properties of each atom are determined by the number of (negatively charged) electrons, which for neutral atoms is equal to the number of (positive) protons so that the total charge is zero. For example, a neutral chlorine atom has 17 protons and 17 electrons, whereas a Cl^- anion has 17 protons and 18 electrons for a total charge of −1.

All atoms of a given element are not necessarily identical, however, as the number of neutrons may vary to form different isotopes, and energy levels may differ forming different nuclear isomers. For example, there are two stable isotopes of

chlorine: 35
17Cl with 35 − 17 = 18 neutrons and 37
17Cl with 37 − 17 = 20 neutrons.

1.7.2 Hydrogen ion

See also: Hydron (chemistry)
In chemistry, the term proton refers to the hydrogen ion, H+
. Since the atomic number of hydrogen is 1, a hydrogen ion has no electrons and corresponds to a bare nucleus, consisting of a proton (and 0 neutrons for the most abundant isotope *protium* 1
1H). The proton is a "bare charge" with only about **1/64,000** of the radius of a hydrogen atom, and so is extremely reactive chemically. The free proton, thus, has an extremely short lifetime in chemical systems such as liquids and it reacts immediately with the electron cloud of any available molecule. In aqueous solution, it forms the hydronium ion, H_3O^+, which in turn is further solvated by water molecules in clusters such as $[H_5O_2]^+$ and $[H_9O_4]^+$.[34]

The transfer of H+
in an acid–base reaction is usually referred to as "proton transfer". The acid is referred to as a proton donor and the base as a proton acceptor. Likewise, biochemical terms such as proton pump and proton channel refer to the movement of hydrated H+
ions.

The ion produced by removing the electron from a deuterium atom is known as a deuteron, not a proton. Likewise, removing an electron from a tritium atom produces a triton.

1.7.3 Proton nuclear magnetic resonance (NMR)

Also in chemistry, the term "proton NMR" refers to the observation of hydrogen-1 nuclei in (mostly organic) molecules by nuclear magnetic resonance. This method uses the spin of the proton, which has the value one-half. The name refers to examination of protons as they occur in protium (hydrogen-1 atoms) in compounds, and does not imply that free protons exist in the compound being studied.

1.8 Human exposure

Main article: Effect of spaceflight on the human body

The Apollo Lunar Surface Experiments Packages (ALSEP) determined that more than 95% of the particles in the solar wind are electrons and protons, in approximately equal numbers.[35][36]

> Because the Solar Wind Spectrometer made continuous measurements, it was possible to measure how the Earth's magnetic field affects arriving solar wind particles. For about two-thirds of each orbit, the Moon is outside of the Earth's magnetic field. At these times, a typical proton density was 10 to 20 per cubic centimeter, with most protons having velocities between 400 and 650 kilometers per second. For about five days of each month, the Moon is inside the Earth's geomagnetic tail, and typically no solar wind particles were detectable. For the remainder of each lunar orbit, the Moon is in a transitional region known as the magnetosheath, where the Earth's magnetic field affects the solar wind but does not completely exclude it. In this region, the particle flux is reduced, with typical proton velocities of 250 to 450 kilometers per second. During the lunar night, the spectrometer was shielded from the solar wind by the Moon and no solar wind particles were measured.[35]

Protons also occur in from extrasolar origin in space, from galactic cosmic rays, where they make up about 90% of the total particle flux. These protons often have higher energy than solar wind protons, but their intensity is far more uniform and less variable than protons coming from the Sun, the production of which is heavily affected by solar proton events such as coronal mass ejections.

Research has been performed on the dose-rate effects of protons, as typically found in space travel, on human health.[36][37] To be more specific, there are hopes to identify what specific chromosomes are damaged, and to define the damage, during cancer development from proton exposure.[36] Another study looks into determining "the effects of exposure to proton irradiation on neurochemical and behavioral endpoints, including dopaminergic functioning, amphetamine-induced conditioned taste aversion learning, and spatial learning and memory as measured by the Morris water maze.[37] Electrical charging of a spacecraft due to interplanetary proton bombardment has also been proposed for study.[38] There are many more studies that pertain to space travel, including galactic cosmic rays and their possible health effects, and solar proton event exposure.

The American Biostack and Soviet Biorack space travel experiments have demonstrated the severity of molecular damage induced by heavy ions on micro organisms including Artemia cysts.[39]

1.9   Antiproton

Main article: Antiproton

CPT-symmetry puts strong constraints on the relative properties of particles and antiparticles and, therefore, is open to stringent tests. For example, the charges of the proton and antiproton must sum to exactly zero. This equality has been tested to one part in 10^8. The equality of their masses has also been tested to better than one part in 10^8. By holding antiprotons in a Penning trap, the equality of the charge to mass ratio of the proton and the antiproton has been tested to one part in 6×10^9.[40] The magnetic moment of the antiproton has been measured with error of 8×10^{-3} nuclear Bohr magnetons, and is found to be equal and opposite to that of the proton.

1.10   See also

- Fermion field

- Hydrogen

- Hydron (chemistry)

- List of particles

- Proton-proton chain reaction

- Quark model

- Proton spin crisis

1.11   References

[1] Mohr, P.J.; Taylor, B.N. and Newell, D.B. (2011), "The 2010 CODATA Recommended Values of the Fundamental Physical Constants", National Institute of Standards and Technology, Gaithersburg, MD, US.

[2] Cho, Adiran (2 April 2010). "Mass of the Common Quark Finally Nailed Down". *http://news.sciencemag.org". American Association for the Advancement of Science. Retrieved 27 September 2014.*

[3] "Proton size puzzle reinforced!". Paul Shearer Institute. 25 January 2013.

[4] Adair, R.K. (1989). *The Great Design: Particles, Fields, and Creation.* Oxford University Press. p. 214.

[5]Cottingham,W.N.;Greenwood,D.A. (1986).*An Introduction to Nuclear Physics.*Cambridge University Press.ISBN97805216573.

[6] Basdevant, J.-L.; Rich, J.; M. Spiro (2005). *Fundamentals in Nuclear Physics*. Springer. p. 155. ISBN 0-387-01672-4.

[7] Department of Chemistry and Biochemistry UCLA Eric R. Scerri Lecturer. *The Periodic Table : Its Story and Its Significance: Its Story and Its Significance*. Oxford University Press. ISBN 978-0-19-534567-4.

[8] Petrucci, R.H.; Harwood, W.S.; Herring, F.G. (2002). *General Chemistry* (8th ed.). p. 41.

[9] See meeting report and announcement

[10] Romer A(1997). "Proton or prouton?Rutherford and the depths of the atom".*Amer.J.Phys.***65**(8):707.Bibcode:1997AmJPh. doi:10.1119/1.18640.

[11] Rutherford reported acceptance by the *British Association* in a footnote to Masson, O. (1921). "XXIV.The constitution of atoms". *Philosophical Magazine Series 6* **41** (242): 281. doi:10.1080/14786442108636219.

[12] Pais, A. (1986) *Inward Bound*, Oxford Press, ISBN 0198519974, p. 296. Pais believed the first science literature use of the word *proton* occurs in "Physics at the British Association". *Nature* **106** (2663): 357. 1920. doi:10.1038/106357a0.

[13] Buccella, F.; Miele, G.; Rosa, L.; Santorelli, P.; Tuzi, T. (1989). "An upper limit for the proton lifetime in SO(10)". *Physics Letters B* **233**: 178. doi:10.1016/0370-2693(89)90637-0.

[14] Lee, D. G.; Mohapatra, R.; Parida, M.; Rani, M. (1995). "Predictions for the proton lifetime in minimal nonsupersymmetric SO(10) models: An update". *Physical Review D* **51**: 229. arXiv:hep-ph/9404238. doi:10.1103/PhysRevD.51.229.

[15] "Proton lifetime is longer than 1034 years". Kamioka Observatory. November 2009.

[16] Nishino, H.; Clark, S.; Abe, K.; Hayato, Y.; Iida, T.; Ikeda, M.; Kameda, J.; Kobayashi, K.; Koshio, Y.; Miura, M.; Moriyama, S.; Nakahata, M.; Nakayama, S.; Obayashi, Y.; Ogawa, H.; Sekiya, H.; Shiozawa, M.; Suzuki, Y.; Takeda, A.; Takenaga, Y.; Takeuchi, Y.; Ueno, K.; Ueshima, K.; Watanabe, H.; Yamada, S.; Hazama, S.; Higuchi, I.; Ishihara, C.; Kajita, T. et al. (2009). "Search for Proton Decay via p→e$^+$π^0 and p→μ$^+$π^0 in a Large Water Cherenkov Detector". *Physical Review Letters* **102** (14). arXiv:0903.0676. Bibcode:2009PhRvL.102n1801N. doi:10.1103/PhysRevLett.102.141801.

[17] Ahmed, S.; Anthony, A.; Beier, E.; Bellerive, A.; Biller, S.; Boger, J.; Boulay, M.; Bowler, M.; Bowles, T.; Brice, S.; Bullard, T.; Chan, Y.; Chen, M.; Chen, X.; Cleveland, B.; Cox, G.; Dai, X.; Dalnoki-Veress, F.; Doe, P.; Dosanjh, R.; Doucas, G.; Dragowsky, M.; Duba, C.; Duncan, F.; Dunford, M.; Dunmore, J.; Earle, E.; Elliott, S.; Evans, H. et al. (2004). "Constraints on Nucleon Decay via Invisible Modes from the Sudbury Neutrino Observatory". *Physical Review Letters* **92** (10). arXiv:hep-ex/0310030. Bibcode:2004PhRvL..92j2004A. doi:10.1103/PhysRevLett.92.102004. PMID 15089201.

[18] Watson, A. (2004). *The Quantum Quark*. Cambridge University Press. pp. 285–286. ISBN 0-521-82907-0.

[19] Timothy Paul Smith (2003). *Hidden Worlds: Hunting for Quarks in Ordinary Matter*. Princeton University Press. ISBN 0-691-05773-7.

[20] Weise, W.; Green, A.M. (1984). *Quarks and Nuclei*. World Scientific. pp. 65–66. ISBN 9971-966-61-1.

[21] Ball, Philip (Nov 20, 2008). "Nuclear masses calculated from scratch". Nature. doi:10.1038/news.2008.1246. Retrieved Aug 27, 2014.

[22] Reynolds, Mark (Apr 2009). "Calculating the Mass of a Proton". *CNRS international magazine* (CNRS) (13). ISSN 2270-5317. Retrieved Aug 27, 2014.

[23] See this news report and links

[24] Durr, S.; Fodor, Z.; Frison, J.; Hoelbling, C.; Hoffmann, R.; Katz, S. D.; Krieg, S.; Kurth, T.; Lellouch, L.; Lippert, T.; Szabo, K. K.; Vulvert, G. (2008). "Ab Initio Determination of Light Hadron Masses". *Science* **322** (5905): 1224–7. arXiv:0906.3599. doi:10.1126/science.1163233. PMID 19023076.

[25] Perdrisat, C. F.; Punjabi, V.; Vanderhaeghen, M. (2007). "Nucleon electromagnetic form factors". *Progress in Particle and Nuclear Physics* **59** (2): 694. arXiv:hep-ph/0612014. Bibcode:2007PrPNP..59..694P. doi:10.1016/j.ppnp.2007.05.001.

[26] Boffi, Sigfrido; Pasquini, Barbara (2007). "Generalized parton distributions and the structure of the nucleon". *Rivista del Nuovo Cimento* **30**. arXiv:0711.2625. Bibcode:2007NCimR..30..387B. doi:10.1393/ncr/i2007-10025-7.

[27] Joshua, Erlich (December 2008). "Recent Results in AdS/QCD". *Proceedings, 8th Conference on Quark Confinement and the Hadron Spectrum, September 1–6, 2008, Mainz, Germany*. arXiv:0812.4976.

[28] Pietro, Colangelo; Alex, Khodjamirian (October 2000). "QCD Sum Rules, a Modern Perspective". In Shifman, M. *At the Frontier of Particle Physics / Handbook of QCD*. World Scientific. arXiv:hep-ph/0010175.

[29] Pohl, R.; Antognini, A.; Nez, F. O.; Amaro, F. D.; Biraben, F. O.; Cardoso, J. O. M. R.; Covita, D. S.; Dax, A.; Dhawan, S.; Fernandes, L. M. P.; Giesen, A.; Graf, T.; Hänsch, T. W.; Indelicato, P.; Julien, L.; Kao, C. Y.; Knowles, P.; Le Bigot, E. O.; Liu, Y. W.; Lopes, J. A. M.; Ludhova, L.; Monteiro, C. M. B.; Mulhauser, F. O.; Nebel, T.; Rabinowitz, P.; Dos Santos, J. M. F.; Schaller, L. A.; Schuhmann, K.; Schwob, C. et al. (2010). "The size of the proton". *Nature* **466** (7303): 213–6. doi:10.1038/nature09250. PMID 20613837.

[30] Antognini, A.; Nez, F.; Schuhmann, K.; Amaro, F. D.; Biraben, F.; Cardoso, J. M. R.; Covita, D. S.; Dax, A.; Dhawan, S.; Diepold, M.; Fernandes, L. M. P.; Giesen, A.; Gouvea, A. L.; Graf, T.; Hänsch, T. W.; Indelicato, P.; Julien, L.; Kao, C. -Y.; Knowles, P.; Kottmann, F.; Le Bigot, E. -O.; Liu, Y. -W.; Lopes, J. A. M.; Ludhova, L.; Monteiro, C. M. B.; Mulhauser, F.; Nebel, T.; Rabinowitz, P.; Dos Santos, J. M. F.; Schaller, L. A. (2013). "Proton Structure from the Measurement of 2S-2P Transition Frequencies of Muonic Hydrogen". *Science* **339** (6118): 417–420. doi:10.1126/science.1230016. PMID 23349284.

[31] Researchers Observes Unexpectedly Small Proton Radius in a Precision Experiment. *Azonano*. July 9, 2010

[32] "The Proton Just Got Smaller". *Photonics.Com*. 12 July 2010. Retrieved 2010-07-19.

[33] Carlson, Carl E. (February 19, 2015), *The Proton Radius Puzzle*, arXiv:1502.05314

[34] Headrick, J.M.; Diken, E.G.; Walters, R. S.; Hammer, N. I.; Christie, R.A.; Cui, J.; Myshakin, E.M.; Duncan, M.A.; Johnson, M.A.; Jordan, K.D. (2005). "Spectral Signatures of Hydrated Proton Vibrations in Water Clusters". *Science* **308** (5729): 1765–69. Bibcode:2005Sci...308.1765H. doi:10.1126/science.1113094. PMID 15961665.

[35] "Apollo 11 Mission". Lunar and Planetary Institute. 2009. Retrieved 2009-06-12.

[36] "Space Travel and Cancer Linked? Stony Brook Researcher Secures NASA Grant to Study Effects of Space Radiation". Brookhaven National Laboratory. 12 December 2007. Retrieved 2009-06-12.

[37] Shukitt-Hale, B.; Szprengiel, A.; Pluhar, J.; Rabin, B.M.; Joseph, J.A. "The effects of proton exposure on neurochemistry and behavior". Elsevier/COSPAR. Retrieved 2009-06-12.

[38] Green, N.W.; Frederickson, A.R. "A Study of Spacecraft Charging due to Exposure to Interplanetary Protons" (PDF). Jet Propulsion Laboratory. Retrieved 2009-06-12.

[39] Planel, H. (2004). *Space and life: an introduction to space biology and medicine*. CRC Press. pp. 135–138. ISBN 0-415-31759-2.

[40] Gabrielse, G. (2006). "Antiproton mass measurements". *International Journal of Mass Spectrometry* **251** (2–3): 273–280. Bibcode:2006IJMSp.251..273G. doi:10.1016/j.ijms.2006.02.013.

1.12 External links

- Particle Data Group

- Large Hadron Collider

- Eaves, Laurence; Copeland, Ed; Padilla, Antonio (Tony) (2010). "The shrinking proton". *Sixty Symbols*. Brady Haran for the University of Nottingham.

Ernest Rutherford at the first Solvay Conference, 1911

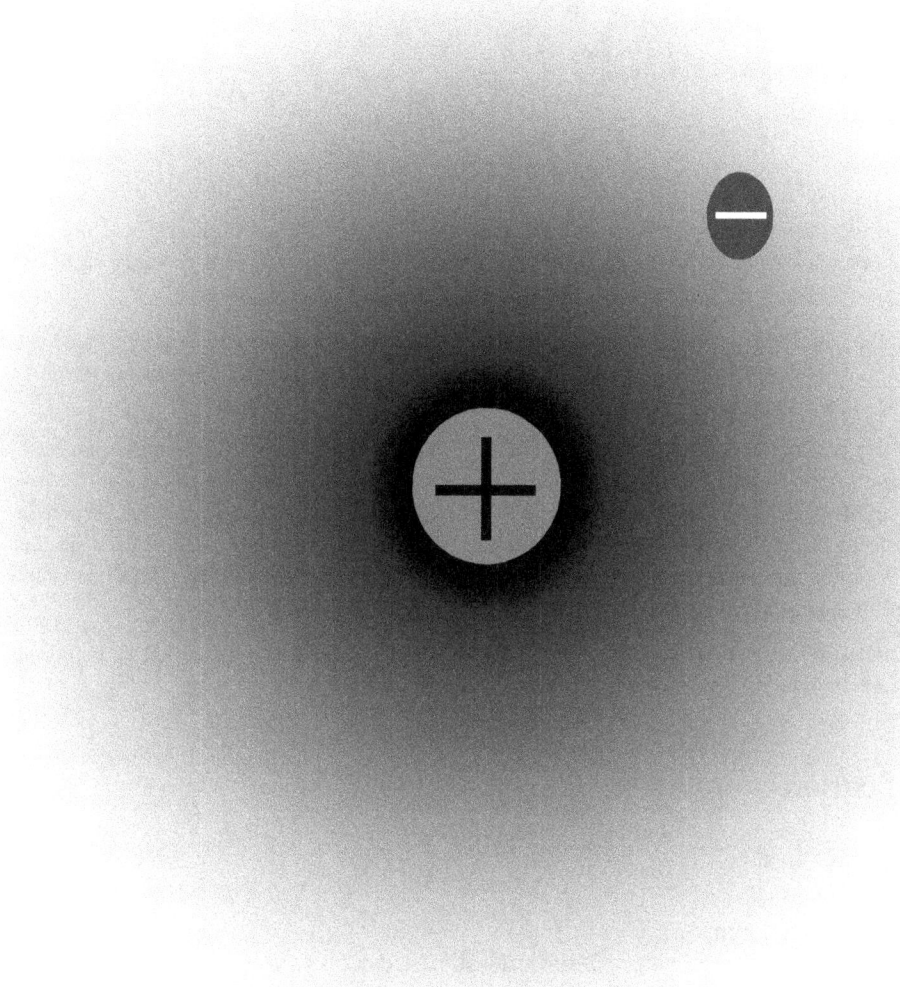

Protium, the most common isotope of hydrogen, consists of one proton and one electron (it has no neutrons). The term "hydrogen ion"
(H+
) implies that that H-atom has lost its one electron, causing only a proton to remain. Thus, in chemistry, the terms "proton" and "hydrogen
ion" (for the protium isotope) are used synonymously

Chapter 2

Subatomic particle

In the physical sciences, **subatomic particles** are particles much smaller than atoms.[1] There are two types of subatomic particles: elementary particles, which according to current theories are not made of other particles; and *composite* particles.[2] Particle physics and nuclear physics study these particles and how they interact.[3]

In particle physics, the concept of a particle is one of several concepts inherited from classical physics. But it also reflects the modern understanding that at the quantum scale matter and energy behave very differently from what much of everyday experience would lead us to expect.

The idea of a particle underwent serious rethinking when experiments showed that light could behave like a stream of particles (called photons) as well as exhibit wave-like properties. This led to the new concept of wave–particle duality to reflect that quantum-scale "particles" behave like both particles and waves (also known as wavicles). Another new concept, the uncertainty principle, states that some of their properties taken together, such as their simultaneous position and momentum, cannot be measured exactly.[4] In more recent times, wave–particle duality has been shown to apply not only to photons but to increasingly massive particles as well.[5]

Interactions of particles in the framework of quantum field theory are understood as creation and annihilation of *quanta* of corresponding fundamental interactions. This blends particle physics with field theory.

2.1 Classification

2.1.1 By statistics

Main article: Spin–statistics theorem
 Any subatomic particle, like any particle in the 3-dimensional space that obeys laws of quantum mechanics, can be either a boson (an integer spin) or a fermion (a half-integer spin).

2.1.2 By composition

The elementary particles of the Standard Model include:[6]

- Six "flavors" of quarks: up, down, bottom, top, strange, and charm;

- Six types of leptons: electron, electron neutrino, muon, muon neutrino, tau, tau neutrino;

- Twelve gauge bosons (force carriers): the photon of electromagnetism, the three W and Z bosons of the weak force, and the eight gluons of the strong force;

- The Higgs boson.

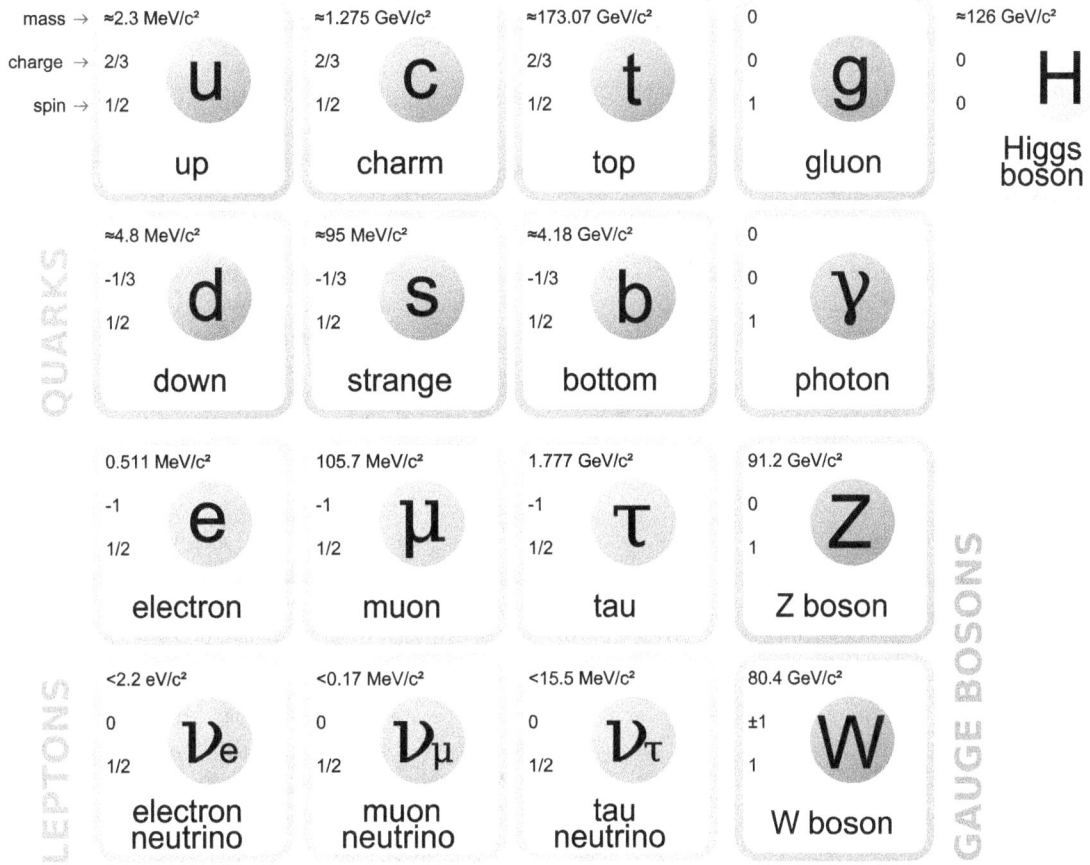

The Standard Model classification of particles

Various extensions of the Standard Model predict the existence of an elementary graviton particle and many other elementary particles.

Composite subatomic particles (such as protons or atomic nuclei) are bound states of two or more elementary particles. For example, a proton is made of two up quarks and one down quark, while the atomic nucleus of helium-4 is composed of two protons and two neutrons. The neutron is made of two down quarks and one up quark. Composite particles include all hadrons: these include baryons (such as protons and neutrons) and mesons (such as pions and kaons).

2.1.3 By mass

In special relativity, the energy of a particle at rest equals its mass times the speed of light squared ($E = mc^2$). That is, mass can be expressed in terms of energy and vice versa. If a particle has a frame of reference where it lies at rest, then it has a positive rest mass and is referred to as *massive*.

All composite particles are massive. Baryons (meaning "heavy") tend to have greater mass than mesons (meaning "intermediate"), which in turn tend to be heavier than leptons (meaning "lightweight"), but the heaviest lepton (the tau particle) is heavier than the two lightest flavours of baryons (nucleons). It is also certain that any particle with an electric charge is massive.

All massless particles (particles whose invariant mass is zero) are elementary. These include the photon and gluon, although the latter cannot be isolated.

The question of the masses of neutrinos is uncertain.

2.2 Other properties

Through the work of Albert Einstein, Louis de Broglie, and many others, current scientific theory holds that *all* particles also have a wave nature.[7] This has been verified not only for elementary particles but also for compound particles like atoms and even molecules. In fact, according to traditional formulations of non-relativistic quantum mechanics, wave–particle duality applies to all objects, even macroscopic ones; although the wave properties of macroscopic objects cannot be detected due to their small wavelengths.[8]

Interactions between particles have been scrutinized for many centuries, and a few simple laws underpin how particles behave in collisions and interactions. The most fundamental of these are the laws of conservation of energy and conservation of momentum, which let us make calculations of particle interactions on scales of magnitude that range from stars to quarks.[9] These are the prerequisite basics of Newtonian mechanics, a series of statements and equations in *Philosophiae Naturalis Principia Mathematica*, originally published in 1687.

2.3 Dividing an atom

The negatively charged electron has a mass equal to $^1/_{1836}$ of that of a hydrogen atom. The remainder of the hydrogen atom's mass comes from the positively charged proton. The atomic number of an element is the number of protons in its nucleus. Neutrons are neutral particles having a mass slightly greater than that of the proton. Different isotopes of the same element contain the same number of protons but differing numbers of neutrons. The mass number of an isotope is the total number of nucleons (neutrons and protons collectively).

Chemistry concerns itself with how electron sharing binds atoms into structures such as crystals and molecules. Nuclear physics deals with how protons and neutrons arrange themselves in nuclei. The study of subatomic particles, atoms and molecules, and their structure and interactions, requires quantum mechanics. Analyzing processes that change the numbers and types of particles requires quantum field theory. The study of subatomic particles *per se* is called particle physics. The term *high-energy physics* is nearly synonymous to "particle physics" since creation of particles requires high energies: it occurs only as a result of cosmic rays, or in particle accelerators. Particle phenomenology systematizes the knowledge about subatomic particles obtained from these experiments.

2.4 History

Main articles: History of subatomic physics and Timeline of particle discoveries

The term "*subatomic* particle" is largely a retronym of 1960s made to distinguish a big number of baryons and mesons (that comprise hadrons) from particles that are now thought to be truly elementary. Before that hadrons were usually classified as "elementary" because their composition was unknown.

A list of important discoveries follows:

2.5 See also

- *Atom: Journey Across the Subatomic Cosmos* (book)

- *Atom: An Odyssey from the Big Bang to Life on Earth...and Beyond* (book)

- CPT invariance

- Dark Matter

- Hot spot effect in subatomic physics

- List of fictional elements, materials, isotopes and atomic particles

- List of particles

- Poincaré symmetry

- Ylem

2.6   References

[1] "Subatomic particles". NTD. Retrieved 5 June 2012.

[2]Bolonkin,Alexander(2011).*Universe,Human Immortality and Future Human Evaluation*.Elsevier.p.25.ISBN9780124158.

[3] Fritzsch, Harald (2005). *Elementary Particles*. World Scientific. pp. 11–20. ISBN 978-981-256-141-1.

[4] Heisenberg, W. (1927), "Über den anschaulichen Inhalt der quantentheoretischen Kinematik und Mechanik", *Zeitschrift für Physik* (in German) **43** (3–4): 172–198, Bibcode:1927ZPhy...43..172H, doi:10.1007/BF01397280.

[5] Arndt, Markus; Nairz, Olaf; Vos-Andreae, Julian; Keller, Claudia; Van Der Zouw, Gerbrand; Zeilinger, Anton (2000). "Wave-particle duality of C60 molecules". *Nature* **401** (6754): 680–682. Bibcode:1999Natur.401..680A. doi:10.1038/44348. PMID 18494170.

[6] Cottingham, W. N.; Greenwood, D. A. (2007). *An introduction to the standard model of particle physics*. Cambridge University Press. p. 1. ISBN 978-0-521-85249-4.

[7] Walter Greiner (2001). *Quantum Mechanics: An Introduction*. Springer. p. 29. ISBN 3-540-67458-6.

[8] R. Eisberg & R. Resnick (1985). *Quantum Physics of Atoms, Molecules, Solids, Nuclei, and Particles* (2nd ed.). John Wiley & Sons. pp. 59–60. ISBN 0-471-87373-X. For both large and small wavelengths, both matter and radiation have both particle and wave aspects. [...] But the wave aspects of their motion become more difficult to observe as their wavelengths become shorter. [...] For ordinary macroscopic particles the mass is so large that the momentum is always sufficiently large to make the de Broglie wavelength small enough to be beyond the range of experimental detection, and classical mechanics reigns supreme.

[9] Isaac Newton (1687). Newton's Laws of Motion (*Philosophiae Naturalis Principia Mathematica*)

[10] Klemperer, Otto (1959). *Electron Physics: The Physics of the Free Electron*. Academic Press.

[11] Some sources such as The Strange Quark indicate 1947.

[12] http://press.web.cern.ch/press-releases/2014/06/cern-experiments-report-new-higgs-boson-measurements

2.7   Further reading

General readers

- Feynman, R.P. & Weinberg, S. (1987). *Elementary Particles and the Laws of Physics: The 1986 Dirac Memorial Lectures*. Cambridge Univ. Press.

- Brian Greene (1999). *The Elegant Universe*. W.W. Norton & Company. ISBN 0-393-05858-1.

- Oerter, Robert (2006). *The Theory of Almost Everything: The Standard Model, the Unsung Triumph of Modern Physics*. Plume.

- Schumm, Bruce A. (2004). *Deep Down Things: The Breathtaking Beauty of Particle Physics*. Johns Hopkins University Press. ISBN 0-8018-7971-X.

- Martinus Veltman (2003). *Facts and Mysteries in Elementary Particle Physics*. World Scientific. ISBN 981-238-149-X.

Textbooks

- Coughlan, G. D., J. E. Dodd, and B. M. Gripaios (2006). *The Ideas of Particle Physics: An Introduction for Scientists*, 3rd ed. Cambridge Univ. Press. An undergraduate text for those not majoring in physics.

- Griffiths, David J. (1987). *Introduction to Elementary Particles*. Wiley, John & Sons, Inc. ISBN 0-471-60386-4.

- Kane, Gordon L. (1987). *Modern Elementary Particle Physics*. Perseus Books. ISBN 0-201-11749-5.

2.8   External links

- particleadventure.org: The Standard Model.

- cpepweb.org: Particle chart.

- University of California: Particle Data Group.

- Annotated Physics Encyclopædia: Quantum Field Theory.

- Jose Galvez: Chapter 1 Electrodynamics (pdf).

Chapter 3

Elementary particle

This article is about the physics concept. For the novel, see The Elementary Particles.
 In particle physics, an **elementary particle** or **fundamental particle** is a particle whose substructure is unknown, thus

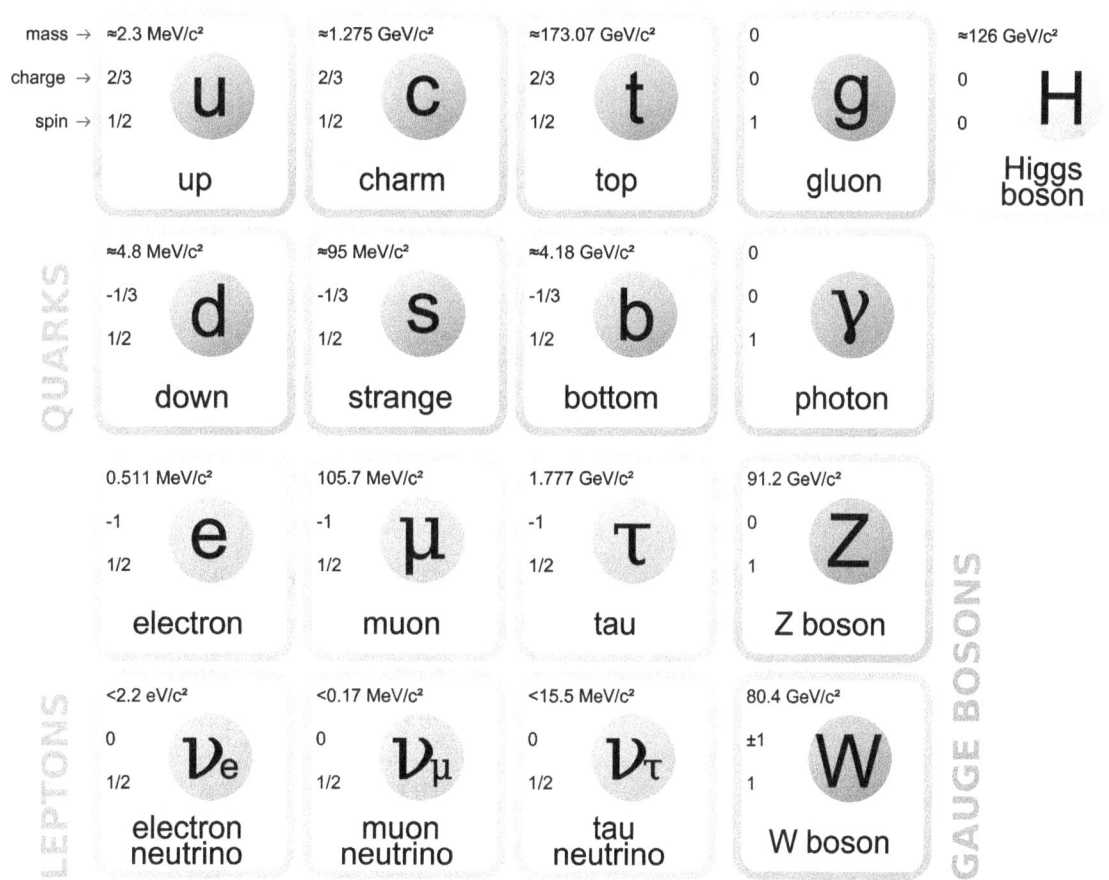

Elementary particles included in the Standard Model

it is unknown whether it is composed of other particles.[1] Known elementary particles include the fundamental fermions (quarks, leptons, antiquarks, and antileptons), which generally are "matter particles" and "antimatter particles", as well as the fundamental bosons (gauge bosons and Higgs boson), which generally are "force particles" that mediate interactions among fermions.[1] A particle containing two or more elementary particles is a *composite particle*.

Everyday matter is composed of atoms, once presumed to be matter's elementary particles—*atom* meaning "indivisible" in Greek—although the atom's existence remained controversial until about 1910, as some leading physicists regarded molecules as mathematical illusions, and matter as ultimately composed of energy.[1][2] Soon, subatomic constituents of the atom were identified. As the 1930s opened, the electron and the proton had been observed, along with the photon, the particle of electromagnetic radiation.[1] At that time, the recent advent of quantum mechanics was radically altering the conception of particles, as a single particle could seemingly span a field as would a wave, a paradox still eluding satisfactory explanation.[3][4][5]

Via quantum theory, protons and neutrons were found to contain quarks—up quarks and down quarks—now considered elementary particles.[1] And within a molecule, the electron's three degrees of freedom (charge, spin, orbital) can separate via wavefunction into three quasiparticles (holon, spinon, orbiton).[6] Yet a free electron—which, not orbiting an atomic nucleus, lacks orbital motion—appears unsplittable and remains regarded as an elementary particle.[6]

Around 1980, an elementary particle's status as indeed elementary—an *ultimate constituent* of substance—was mostly discarded for a more practical outlook,[1] embodied in particle physics' Standard Model, science's most experimentally successful theory.[5][7] Many elaborations upon and theories beyond the Standard Model, including the extremely popular supersymmetry, double the number of elementary particles by hypothesizing that each known particle associates with a "shadow" partner far more massive,[8][9] although all such superpartners remain undiscovered.[7][10] Meanwhile, an elementary boson mediating gravitation—the graviton—remains hypothetical.[1]

3.1 Overview

Main article: Standard Model
See also: Physics beyond the Standard Model
All elementary particles are—depending on their *spin*—either bosons or fermions. These are differentiated via the spin–

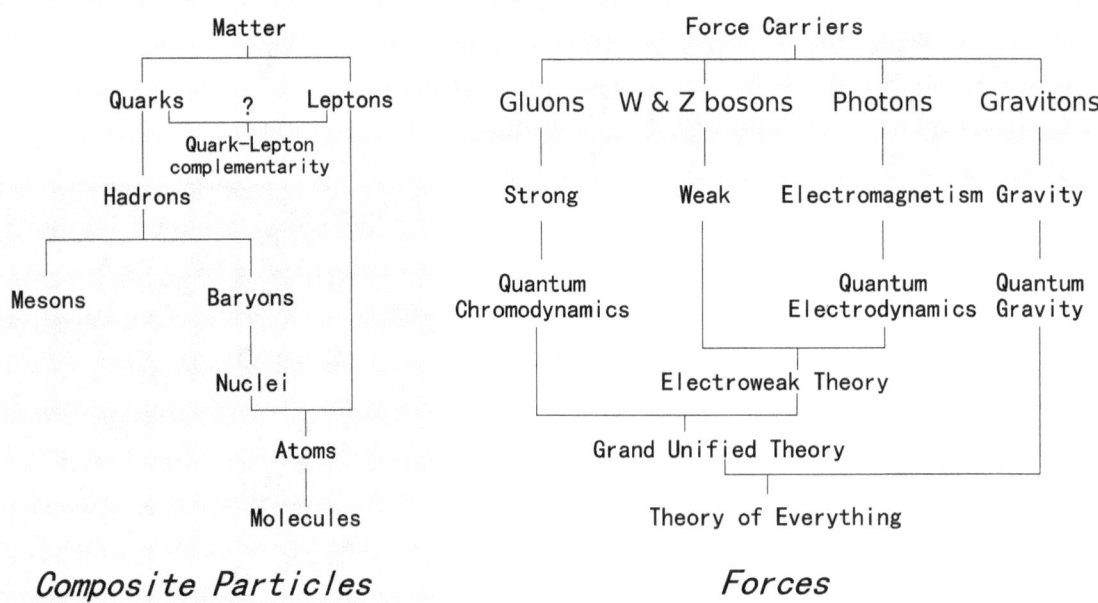

An overview of the various families of elementary and composite particles, and the theories describing their interactions

statistics theorem of quantum statistics. Particles of *half-integer* spin exhibit Fermi–Dirac statistics and are fermions.[1] Particles of *integer* spin, in other words full-integer, exhibit Bose–Einstein statistics and are bosons.[1]

Elementary fermions:

- Matter particles

  - Quarks:

    - up, down
    - charm, strange
    - top, bottom

  - Leptons:

    - electron, electron neutrino (a.k.a., "neutrino")
    - muon, muon neutrino
    - tau, tau neutrino

- Antimatter particles

  - Antiquarks

  - Antileptons

Elementary bosons:

- Force particles (gauge bosons):

  - photon
  - gluon (numbering eight)[1]
  - W^+, W^-, and Z^0 bosons
  - graviton (hypothetical)[1]

- Scalar boson

  - Higgs boson

A particle's mass is quantified in units of energy versus the electron's (electronvolts). Through conversion of energy into mass, any particle can be produced through collision of other particles at high energy,[1][11] although the output particle might not contain the input particles, for instance matter creation from colliding photons. Likewise, the composite fermions protons were collided at nearly light speed to produce a Higgs boson, which elementary boson is far more massive.[11] The most massive elementary particle, the top quark, rapidly decays, but apparently does not contain, lighter particles.

When probed at energies available in experiments, particles exhibit spherical sizes. In operating particle physics' Standard Model, elementary particles are usually represented for predictive utility as point particles, which, as zero-dimensional, lack spatial extension. Though extremely successful, the Standard Model is limited to the microcosm by its omission of gravitation, and has some parameters arbitrarily added but unexplained.[12] Seeking to resolve those shortcomings, string theory posits that elementary particles are ultimately composed of one-dimensional energy strings whose absolute minimum size is the Planck length.

3.2   Common elementary particles

Main article: cosmic abundance of elements

According to the current models of big bang nucleosynthesis, the primordial composition of visible matter of the universe should be about 75% hydrogen and 25% helium-4 (in mass). Neutrons are made up of one up and two down quark, while protons are made of two up and one down quark. Since the other common elementary particles (such as electrons, neutrinos, or weak bosons) are so light or so rare when compared to atomic nuclei, we can neglect their mass contribution to the observable universe's total mass. Therefore, one can conclude that most of the visible mass of the universe consists of protons and neutrons, which, like all baryons, in turn consist of up quarks and down quarks.

Some estimates imply that there are roughly 10^{80} baryons (almost entirely protons and neutrons) in the observable universe.[13][14][15]

The number of protons in the observable universe is called the Eddington number.

In terms of number of particles, some estimates imply that nearly all the matter, excluding dark matter, occurs in neutrinos, and that roughly 10^{86} elementary particles of matter exist in the visible universe, mostly neutrinos.[15] Other estimates imply that roughly 10^{97} elementary particles exist in the visible universe (not including dark matter), mostly photons, gravitons, and other massless force carriers.[15]

3.3 Standard Model

Main article: Standard Model
The Standard Model of particle physics contains 12 flavors of elementary fermions, plus their corresponding antiparticles, as well as elementary bosons that mediate the forces and the Higgs boson, which was reported on July 4, 2012, as having been likely detected by the two main experiments at the LHC (ATLAS and CMS). However, the Standard Model is widely considered to be a provisional theory rather than a truly fundamental one, since it is not known if it is compatible with Einstein's general relativity. There may be hypothetical elementary particles not described by the Standard Model, such as the graviton, the particle that would carry the gravitational force, and sparticles, supersymmetric partners of the ordinary particles.

3.3.1 Fundamental fermions

Main article: Fermion

The 12 fundamental fermionic flavours are divided into three generations of four particles each. Six of the particles are quarks. The remaining six are leptons, three of which are neutrinos, and the remaining three of which have an electric charge of −1: the electron and its two cousins, the muon and the tau.

Antiparticles

Main article: Antimatter

There are also 12 fundamental fermionic antiparticles that correspond to these 12 particles. For example, the antielectron (positron) $e+$ is the electron's antiparticle and has an electric charge of +1.

Quarks

Main article: Quark

Isolated quarks and antiquarks have never been detected, a fact explained by confinement. Every quark carries one of three color charges of the strong interaction; antiquarks similarly carry anticolor. Color-charged particles interact via gluon exchange in the same way that charged particles interact via photon exchange. However, gluons are themselves color-charged, resulting in an amplification of the strong force as color-charged particles are separated. Unlike the electromagnetic force, which diminishes as charged particles separate, color-charged particles feel increasing force.

However, color-charged particles may combine to form color neutral composite particles called hadrons. A quark may pair up with an antiquark: the quark has a color and the antiquark has the corresponding anticolor. The color and anticolor cancel out, forming a color neutral meson. Alternatively, three quarks can exist together, one quark being "red",

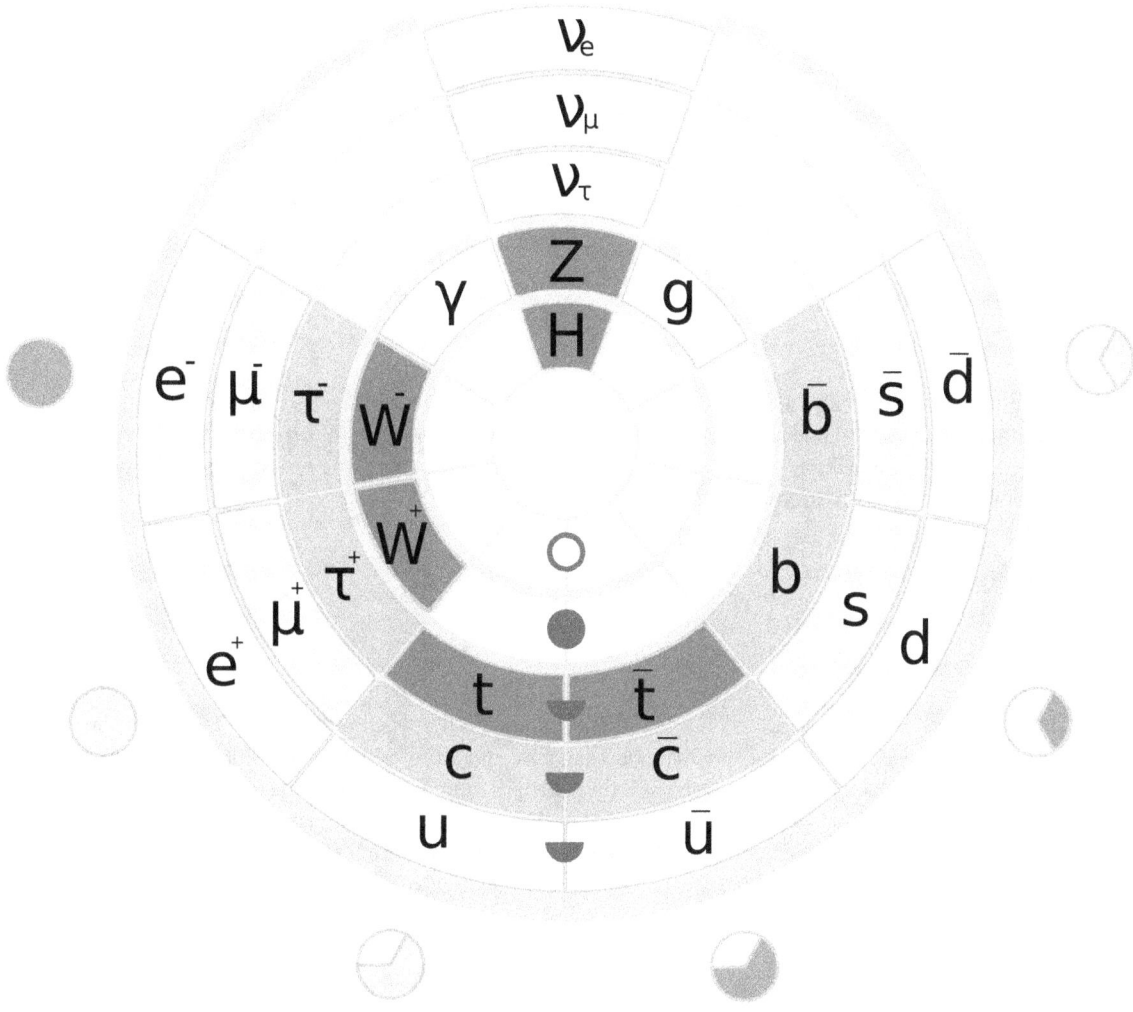

Graphic representation of the standard model. Spin, charge, mass and participation in different force interactions are shown. Click on the image to see the full description

another "blue", another "green". These three colored quarks together form a color-neutral baryon. Symmetrically, three antiquarks with the colors "antired", "antiblue" and "antigreen" can form a color-neutral antibaryon.

Quarks also carry fractional electric charges, but, since they are confined within hadrons whose charges are all integral, fractional charges have never been isolated. Note that quarks have electric charges of either +2/3 or −1/3, whereas antiquarks have corresponding electric charges of either −2/3 or +1/3.

Evidence for the existence of quarks comes from deep inelastic scattering: firing electrons at nuclei to determine the distribution of charge within nucleons (which are baryons). If the charge is uniform, the electric field around the proton should be uniform and the electron should scatter elastically. Low-energy electrons do scatter in this way, but, above a particular energy, the protons deflect some electrons through large angles. The recoiling electron has much less energy and a jet of particles is emitted. This inelastic scattering suggests that the charge in the proton is not uniform but split among smaller charged particles: quarks.

3.3.2 Fundamental bosons

Main article: Boson

In the Standard Model, vector (spin−1) bosons (gluons, photons, and the W and Z bosons) mediate forces, whereas the Higgs boson (spin-0) is responsible for the intrinsic mass of particles. Bosons differ from fermions in the fact that multiple bosons can occupy the same quantum state (Pauli exclusion principle). Also, bosons can be either elementary, like photons, or a combination, like mesons. The spin of bosons are integers instead of half integers.

Gluons

Main article: Gluon

Gluons mediate the strong interaction, which join quarks and thereby form hadrons, which are either baryons (three quarks) or mesons (one quark and one antiquark). Protons and neutrons are baryons, joined by gluons to form the atomic nucleus. Like quarks, gluons exhibit colour and anticolour—unrelated to the concept of visual color—sometimes in combinations, altogether eight variations of gluons.

Electroweak bosons

Main articles: W and Z bosons and Photon

There are three weak gauge bosons: W^+, W^-, and Z^0; these mediate the weak interaction. The W bosons are known for their mediation in nuclear decay. The W^- converts a neutron into a proton then decay into an electron and electron antineutrino pair. The Z^0 does not convert charge but rather changes momentum and is the only mechanism for elastically scattering neutrinos. The weak gauge bosons were discovered due to momentum change in electrons from neutrino-Z exchange. The massless photon mediates the electromagnetic interaction. These four gauge bosons form the electroweak interaction among elementary particles.

Higgs boson

Main article: Higgs boson

Although the weak and electromagnetic forces appear quite different to us at everyday energies, the two forces are theorized to unify as a single electroweak force at high energies. This prediction was clearly confirmed by measurements of cross-sections for high-energy electron-proton scattering at the HERA collider at DESY. The differences at low energies is a consequence of the high masses of the W and Z bosons, which in turn are a consequence of the Higgs mechanism. Through the process of spontaneous symmetry breaking, the Higgs selects a special direction in electroweak space that causes three electroweak particles to become very heavy (the weak bosons) and one to remain massless (the photon). On 4 July 2012, after many years of experimentally searching for evidence of its existence, the Higgs boson was announced to have been observed at CERN's Large Hadron Collider. Peter Higgs who first posited the existence of the Higgs boson was present at the announcement.[16] The Higgs boson is believed to have a mass of approximately 125 GeV.[17] The statistical significance of this discovery was reported as 5-sigma, which implies a certainty of roughly 99.99994%. In particle physics, this is the level of significance required to officially label experimental observations as a discovery. Research into the properties of the newly discovered particle continues.

Graviton

Main article: Graviton

The graviton is hypothesized to mediate gravitation, but remains undiscovered and yet is sometimes included in tables of elementary particles.[1] Its spin would be two—thus a boson—and it would lack charge or mass. Besides mediating an extremely feeble force, the graviton would have its own antiparticle and rapidly annihilate, rendering its detection extremely difficult even if it exists.

3.4   Beyond the Standard Model

Although experimental evidence overwhelmingly confirms the predictions derived from the Standard Model, some of its parameters were added arbitrarily, not determined by a particular explanation, which remain mysteries, for instance the hierarchy problem. Theories beyond the Standard Model attempt to resolve these shortcomings.

3.4.1   Grand unification

Main article: Grand Unified Theory

One extension of the Standard Model attempts to combine the electroweak interaction with the strong interaction into a single 'grand unified theory' (GUT). Such a force would be spontaneously broken into the three forces by a Higgs-like mechanism. The most dramatic prediction of grand unification is the existence of X and Y bosons, which cause proton decay. However, the non-observation of proton decay at the Super-Kamiokande neutrino observatory rules out the simplest GUTs, including SU(5) and SO(10).

3.4.2   Supersymmetry

Main article: Supersymmetry

Supersymmetry extends the Standard Model by adding another class of symmetries to the Lagrangian. These symmetries exchange fermionic particles with bosonic ones. Such a symmetry predicts the existence of supersymmetric particles, abbreviated as *sparticles*, which include the sleptons, squarks, neutralinos, and charginos. Each particle in the Standard Model would have a superpartner whose spin differs by 1/2 from the ordinary particle. Due to the breaking of supersymmetry, the sparticles are much heavier than their ordinary counterparts; they are so heavy that existing particle colliders would not be powerful enough to produce them. However, some physicists believe that sparticles will be detected by the Large Hadron Collider at CERN.

3.4.3   String theory

Main article: String theory

String theory is a model of physics where all "particles" that make up matter are composed of strings (measuring at the Planck length) that exist in an 11-dimensional (according to M-theory, the leading version) universe. These strings vibrate at different frequencies that determine mass, electric charge, color charge, and spin. A string can be open (a line) or closed in a loop (a one-dimensional sphere, like a circle). As a string moves through space it sweeps out something called a *world sheet*. String theory predicts 1- to 10-branes (a 1-brane being a string and a 10-brane being a 10-dimensional object) that prevent tears in the "fabric" of space using the uncertainty principle (E.g., the electron orbiting a hydrogen atom has the probability, albeit small, that it could be anywhere else in the universe at any given moment).

String theory proposes that our universe is merely a 4-brane, inside which exist the 3 space dimensions and the 1 time dimension that we observe. The remaining 6 theoretical dimensions either are very tiny and curled up (and too small to be macroscopically accessible) or simply do not/cannot exist in our universe (because they exist in a grander scheme called the "multiverse" outside our known universe).

Some predictions of the string theory include existence of extremely massive counterparts of ordinary particles due to vibrational excitations of the fundamental string and existence of a massless spin-2 particle behaving like the graviton.

3.4.4 Technicolor

Main article: Technicolor (physics)

Technicolor theories try to modify the Standard Model in a minimal way by introducing a new QCD-like interaction. This means one adds a new theory of so-called Techniquarks, interacting via so called Technigluons. The main idea is that the Higgs-Boson is not an elementary particle but a bound state of these objects.

3.4.5 Preon theory

Main article: Preon

According to preon theory there are one or more orders of particles more fundamental than those (or most of those) found in the Standard Model. The most fundamental of these are normally called preons, which is derived from "pre-quarks". In essence, preon theory tries to do for the Standard Model what the Standard Model did for the particle zoo that came before it. Most models assume that almost everything in the Standard Model can be explained in terms of three to half a dozen more fundamental particles and the rules that govern their interactions. Interest in preons has waned since the simplest models were experimentally ruled out in the 1980s.

3.4.6 Acceleron theory

Accelerons are the hypothetical subatomic particles that integrally link the newfound mass of the neutrino and to the dark energy conjectured to be accelerating the expansion of the universe.[18]

In theory, neutrinos are influenced by a new force resulting from their interactions with accelerons. Dark energy results as the universe tries to pull neutrinos apart.[18]

3.5 See also

- Asymptotic freedom

- List of particles

- Physical ontology

- Quantum field theory

- Quantum gravity

- Quantum triviality

- UV fixed point

3.6 Notes

[1] Sylvie Braibant; Giorgio Giacomelli; Maurizio Spurio (2012). *Particles and Fundamental Interactions: An Introduction to Particle Physics* (2nd ed.). Springer. pp. 1–3. ISBN 978-94-007-2463-1.

[2] Ronald Newburgh; Joseph Peidle; Wolfgang Rueckner (2006). "Einstein, Perrin, and the reality of atoms: 1905 revisited" (PDF). *American Journal of Physics*. **74** (6): 478–481. Bibcode:2006AmJPh..74..478N. doi:10.1119/1.2188962.

[3] Friedel Weinert (2004). *The Scientist as Philosopher: Philosophical Consequences of Great Scientific Discoveries*. Springer. p. 43. ISBN 978-3-540-20580-7.

[4] Friedel Weinert (2004). *The Scientist as Philosopher: Philosophical Consequences of Great Scientific Discoveries*. Springer. pp. 57–59. ISBN 978-3-540-20580-7.

[5] Meinard Kuhlmann (24 Jul 2013). "Physicists debate whether the world is made of particles or fields—or something else entirely". *Scientific American*.

[6] Zeeya Merali (18 Apr 2012). "Not-quite-so elementary, my dear electron: Fundamental particle 'splits' into quasiparticles, including the new 'orbiton'". *Nature*. doi:10.1038/nature.2012.10471.

[7] Ian O'Neill (24 Jul 2013). "LHC discovery maims supersymmetry, again". *Discovery News*. Retrieved 2013-08-28.

[8] Particle Data Group. "Unsolved mysteries—supersymmetry". *The Particle Adventure*. Berkeley Lab. Retrieved 2013-08-28.

[9] National Research Council (2006). *Revealing the Hidden Nature of Space and Time: Charting the Course for Elementary Particle Physics*. National Academies Press. p. 68. ISBN 978-0-309-66039-6.

[10] "CERN latest data shows no sign of supersymmetry—yet". *Phys.Org*. 25 Jul 2013. Retrieved 2013-08-28.

[11] Ryan Avent (19 Jul 2012). "The Q&A: Brian Greene—Life after the Higgs". *The Economist*. Retrieved 2013-08-28.

[12] Sylvie Braibant; Giorgio Giacomelli; Maurizio Spurio (2012). *Particles and Fundamental Interactions: An Introduction to Particle Physics* (2nd ed.). Springer. p. 384. ISBN 978-94-007-2463-1.

[13] Frank Heile. "Is the Total Number of Particles in the Universe Stable Over Long Periods of Time?". 2014.

[14] Jared Brooks. "Galaxies and Cosmology". 2014. p. 4, equation 16.

[15] Robert Munafo (24 Jul 2013). "Notable Properties of Specific Numbers". Retrieved 2013-08-28.

[16] Lizzy Davies (4 July 2014). "Higgs boson announcement live: CERN scientists discover subatomic particle". *The Guardian*. Retrieved 2012-07-06.

[17] Lucas Taylor (4 Jul 2014). "Observation of a new particle with a mass of 125 GeV". CMS. Retrieved 2012-07-06.

[18] "New theory links neutrino's slight mass to accelerating Universe expansion". *ScienceDaily*. 28 Jul 2004. Retrieved 2008-06-05.

3.7   Further reading

3.7.1   General readers

- Feynman, R.P. & Weinberg, S. (1987) *Elementary Particles and the Laws of Physics: The 1986 Dirac Memorial Lectures*. Cambridge Univ. Press.

- Ford, Kenneth W. (2005) *The Quantum World*. Harvard Univ. Press.

- Brian Greene (1999). *The Elegant Universe*. W.W.Norton & Company. ISBN 0-393-05858-1.

- John Gribbin (2000) *Q is for Quantum – An Encyclopedia of Particle Physics*. Simon & Schuster. ISBN 0-684-85578-X.

- Oerter, Robert (2006) *The Theory of Almost Everything: The Standard Model, the Unsung Triumph of Modern Physics*. Plume.

- Schumm, Bruce A. (2004) *Deep Down Things: The Breathtaking Beauty of Particle Physics*. Johns Hopkins University Press. ISBN 0-8018-7971-X.

- Martinus Veltman (2003). *Facts and Mysteries in Elementary Particle Physics*. World Scientific. ISBN 981-238-149-X.

- Frank Close (2004). *Particle Physics: A Very Short Introduction*. Oxford: Oxford University Press. ISBN 0-19-280434-0.

- Seiden, Abraham (2005). *Particle Physics – A Comprehensive Introduction*. Addison Wesley. ISBN 0-8053-8736-6.

3.7.2 Textbooks

- Bettini, Alessandro (2008) *Introduction to Elementary Particle Physics*. Cambridge Univ. Press. ISBN 978-0-521-88021-3

- Coughlan, G. D., J. E. Dodd, and B. M. Gripaios (2006) *The Ideas of Particle Physics: An Introduction for Scientists*, 3rd ed. Cambridge Univ. Press. An undergraduate text for those not majoring in physics.

- Griffiths, David J. (1987) *Introduction to Elementary Particles*. John Wiley & Sons. ISBN 0-471-60386-4.

- Kane, Gordon L. (1987). *Modern Elementary Particle Physics*. Perseus Books. ISBN 0-201-11749-5.

- Perkins, Donald H. (2000) *Introduction to High Energy Physics*, 4th ed. Cambridge Univ. Press.

3.8 External links

The most important address about the current experimental and theoretical knowledge about elementary particle physics is the Particle Data Group, where different international institutions collect all experimental data and give short reviews over the contemporary theoretical understanding.

- Particle Data Group

other pages are:

- Greene, Brian, "*Elementary particles*", The Elegant Universe, NOVA (PBS)

- particleadventure.org, a well-made introduction also for non physicists

- CERNCourier: Season of Higgs and melodrama

- Pentaquark information page

- Interactions.org, particle physics news

- Symmetry Magazine, a joint Fermilab/SLAC publication

- "Sized Matter: perception of the extreme unseen", Michigan University project for artistic visualisation of sub-atomic particles

- Elementary Particles made thinkable, an interactive visualisation allowing physical properties to be compared

Chapter 4

List of particles

This is a list of the different types of particles found or believed to exist in the whole of the universe. For individual lists of the different particles, see the list below.

4.1 Elementary particles

Main article: Elementary particle

Elementary particles are particles with no measurable internal structure; that is, they are not composed of other particles. They are the fundamental objects of quantum field theory. Many families and sub-families of elementary particles exist. Elementary particles are classified according to their spin. Fermions have half-integer spin while bosons have integer spin. All the particles of the Standard Model have been experimentally observed, recently including the Higgs boson.[1][2]

4.1.1 Fermions

Main article: Fermion

Fermions are one of the two fundamental classes of particles, the other being bosons. Fermion particles are described by Fermi–Dirac statistics and have quantum numbers described by the Pauli exclusion principle. They include the quarks and leptons, as well as any composite particles consisting of an odd number of these, such as all baryons and many atoms and nuclei.

Fermions have half-integer spin; for all known elementary fermions this is $\frac{1}{2}$. All known fermions, except neutrinos, are also Dirac fermions; that is, each known fermion has its own distinct antiparticle. It is not known whether the neutrino is a Dirac fermion or a Majorana fermion.[3] Fermions are the basic building blocks of all matter. They are classified according to whether they interact via the color force or not. In the Standard Model, there are 12 types of elementary fermions: six quarks and six leptons.

Quarks

Main article: Quark

Quarks are the fundamental constituents of hadrons and interact via the strong interaction. Quarks are the only known carriers of fractional charge, but because they combine in groups of three (baryons) or in groups of two with antiquarks (mesons), only integer charge is observed in nature. Their respective antiparticles are the antiquarks, which are identical except for the fact that they carry the opposite electric charge (for example the up quark carries charge $+^2/_3$, while the up antiquark carries charge $-^2/_3$), color charge, and baryon number. There are six flavors of quarks; the three positively charged quarks are called "up-type quarks" and the three negatively charged quarks are called "down-type quarks".

Leptons

Main article: Leptons

Leptons do not interact via the strong interaction. Their respective antiparticles are the antileptons which are identical, except for the fact that they carry the opposite electric charge and lepton number. The antiparticle of an electron is an antielectron, which is nearly always called a "positron" for historical reasons. There are six leptons in total; the three charged leptons are called "electron-like leptons", while the neutral leptons are called "neutrinos". Neutrinos are known to oscillate, so that neutrinos of definite flavor do not have definite mass, rather they exist in a superposition of mass eigenstates. The hypothetical heavy right-handed neutrino, called a "sterile neutrino", has been left off the list.

4.1.2 Bosons

Main article: Boson

Bosons are one of the two fundamental classes of particles, the other being fermions. Bosons are characterized by Bose–Einstein statistics and all have integer spins. Bosons may be either elementary, like photons and gluons, or composite, like mesons.

The fundamental forces of nature are mediated by gauge bosons, and mass is believed to be created by the Higgs field. According to the Standard Model the elementary bosons are:

The graviton is added to the list although it is not predicted by the Standard Model, but by other theories in the framework of quantum field theory. Furthermore, gravity is non-renormalizable. There are a total of eight independent gluons. The Higgs boson is postulated by the electroweak theory primarily to explain the origin of particle masses. In a process known as the "Higgs mechanism", the Higgs boson and the other gauge bosons in the Standard Model acquire mass via spontaneous symmetry breaking of the SU(2) gauge symmetry. The Minimal Supersymmetric Standard Model (MSSM) predicts several Higgs bosons. A new particle expected to be the Higgs boson was observed at the CERN/LHC on March 14, 2013, around the energy of 126.5GeV with an accuracy of close to five sigma (99.9999%, which is accepted as definitive). The Higgs mechanism giving mass to other particles has not been observed yet.

4.1.3 Hypothetical particles

Supersymmetric theories predict the existence of more particles, none of which have been confirmed experimentally as of 2014:

Note: just as the photon, Z boson and $W^\pm$ bosons are superpositions of the B^0, W^0, W^1, and W^2 fields – the photino, zino, and wino$^\pm$ are superpositions of the bino^0, wino^0, wino^1, and wino^2 by definition.

No matter if one uses the original gauginos or this superpositions as a basis, the only predicted physical particles are neutralinos and charginos as a superposition of them together with the Higgsinos.

Other theories predict the existence of additional bosons:

Mirror particles are predicted by theories that restore parity symmetry.

"Magnetic monopole" is a generic name for particles with non-zero magnetic charge. They are predicted by some GUTs.

"Tachyon" is a generic name for hypothetical particles that travel faster than the speed of light and have an imaginary rest mass.

Preons were suggested as subparticles of quarks and leptons, but modern collider experiments have all but ruled out their existence.

Kaluza–Klein towers of particles are predicted by some models of extra dimensions. The extra-dimensional momentum is manifested as extra mass in four-dimensional spacetime.

4.2   Composite particles

4.2.1   Hadrons

Main article: Hadron

Hadrons are defined as strongly interacting composite particles. Hadrons are either:

- Composite fermions, in which case they are called baryons.
- Composite bosons, in which case they are called mesons.

Quark models, first proposed in 1964 independently by Murray Gell-Mann and George Zweig (who called quarks "aces"), describe the known hadrons as composed of valence quarks and/or antiquarks, tightly bound by the color force, which is mediated by gluons. A "sea" of virtual quark-antiquark pairs is also present in each hadron.

Baryons

See also: List of baryons

Ordinary baryons (composite fermions) contain three valence quarks or three valence antiquarks each.

- Nucleons are the fermionic constituents of normal atomic nuclei:
  - Protons, composed of two up and one down quark (uud)
  - Neutrons, composed of two down and one up quark (ddu)
- Hyperons, such as the Λ, Σ, Ξ, and Ω particles, which contain one or more strange quarks, are short-lived and heavier than nucleons. Although not normally present in atomic nuclei, they can appear in short-lived hypernuclei.
- A number of charmed and bottom baryons have also been observed.

Some hints at the existence of exotic baryons have been found recently; however, negative results have also been reported. Their existence is uncertain.

- Pentaquarks consist of four valence quarks and one valence antiquark.

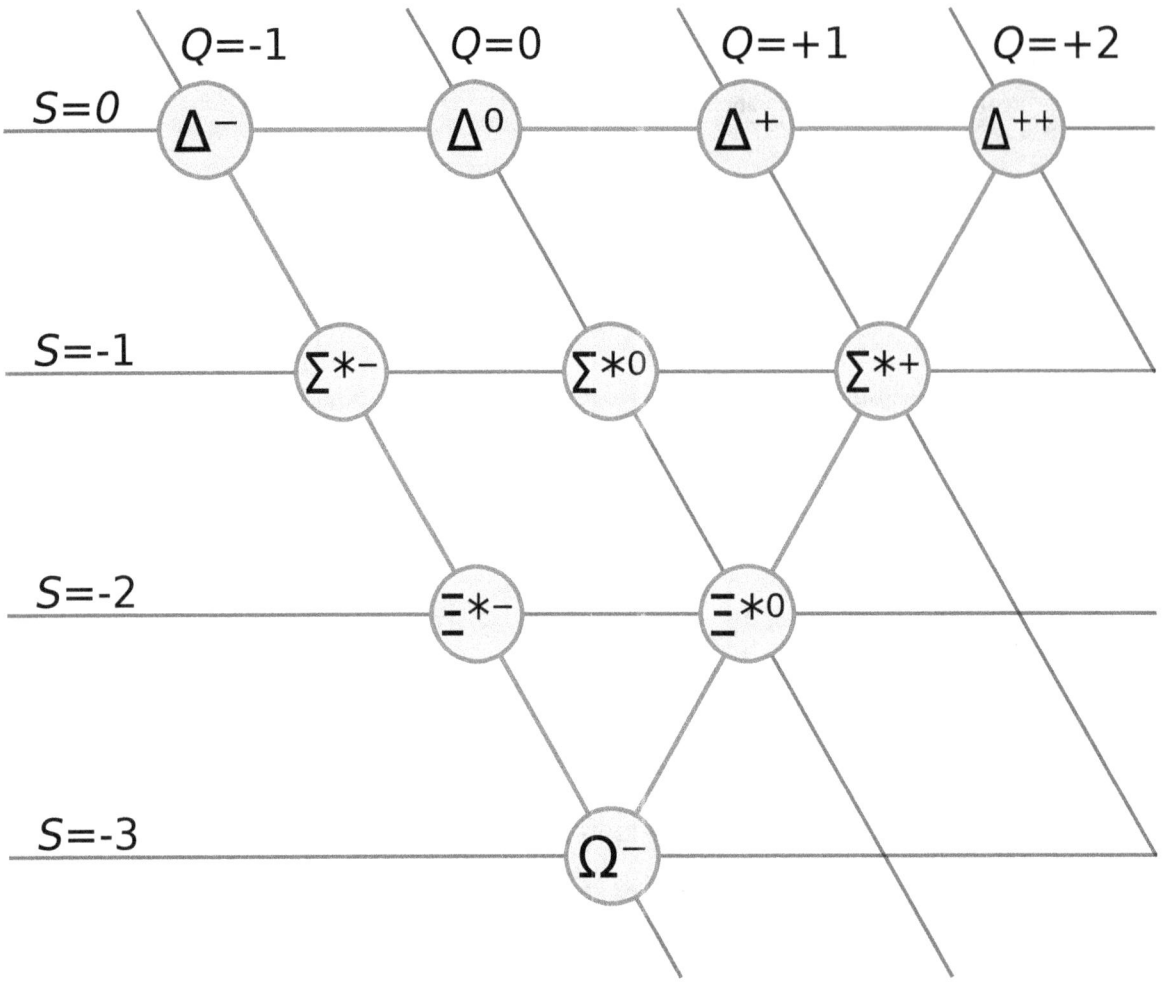

A combination of three u, d or s-quarks with a total spin of $^3\!/_2$ form the so-called "baryon decuplet".

Mesons

See also: List of mesons

Ordinary mesons are made up of a valence quark and a valence antiquark. Because mesons have spin of 0 or 1 and are not themselves elementary particles, they are "composite" bosons. Examples of mesons include the pion, kaon, and the J/ψ. In quantum hydrodynamic models, mesons mediate the residual strong force between nucleons.

At one time or another, positive signatures have been reported for all of the following exotic mesons but their existences have yet to be confirmed.

- A tetraquark consists of two valence quarks and two valence antiquarks;

- A glueball is a bound state of gluons with no valence quarks;

- Hybrid mesons consist of one or more valence quark-antiquark pairs and one or more real gluons.

4.2.2 Atomic nuclei

Atomic nuclei consist of protons and neutrons. Each type of nucleus contains a specific number of protons and a specific

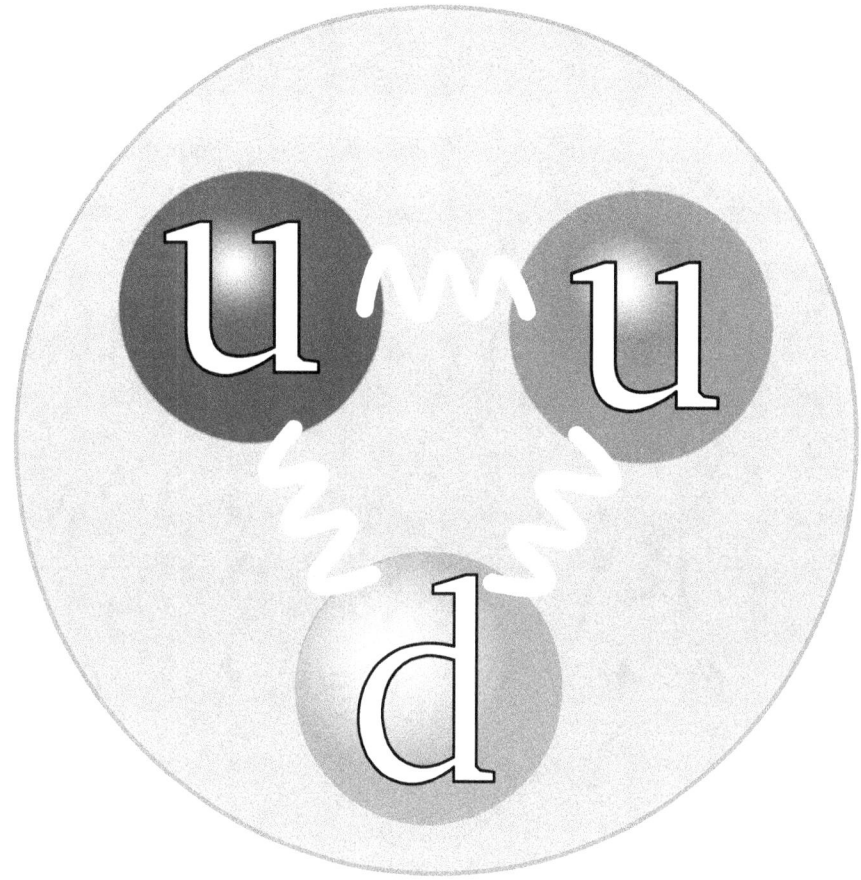

Proton quark structure: 2 up quarks and 1 down quark. The gluon tubes or flux tubes are now known to be Y shaped.

number of neutrons, and is called a "nuclide" or "isotope". Nuclear reactions can change one nuclide into another. See table of nuclides for a complete list of isotopes.

4.2.3   Atoms

Atoms are the smallest neutral particles into which matter can be divided by chemical reactions. An atom consists of a small, heavy nucleus surrounded by a relatively large, light cloud of electrons. Each type of atom corresponds to a specific chemical element. To date, 118 elements have been discovered, while only the elements 1-112,114, and 116 have received official names.

The atomic nucleus consists of protons and neutrons. Protons and neutrons are, in turn, made of quarks.

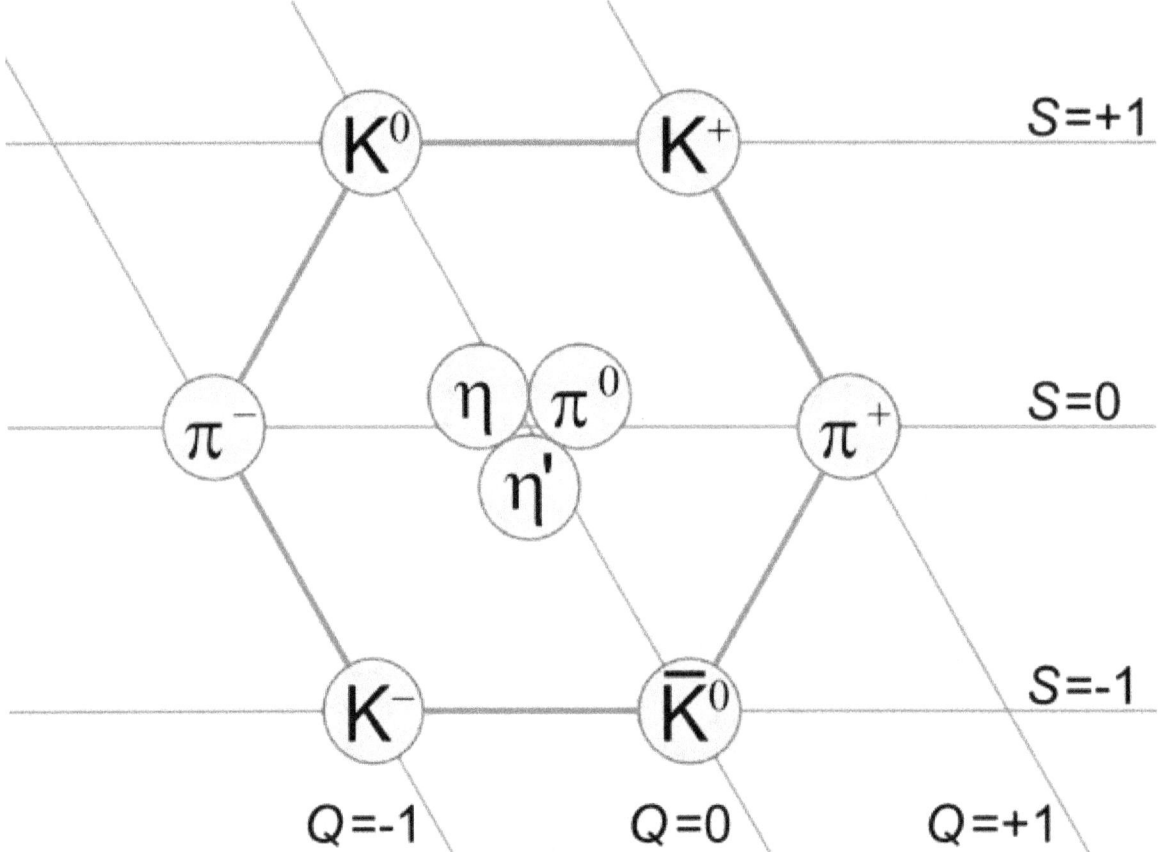

Mesons of spin 0 form a nonet

4.2.4 Molecules

Molecules are the smallest particles into which a non-elemental substance can be divided while maintaining the physical properties of the substance. Each type of molecule corresponds to a specific chemical compound. Molecules are a composite of two or more atoms. See list of compounds for a list of molecules.

4.3 Condensed matter

The field equations of condensed matter physics are remarkably similar to those of high energy particle physics. As a result, much of the theory of particle physics applies to condensed matter physics as well; in particular, there are a selection of field excitations, called quasi-particles, that can be created and explored. These include:

- Phonons are vibrational modes in a crystal lattice.

- Excitons are bound states of an electron and a hole.

- Plasmons are coherent excitations of a plasma.

- Polaritons are mixtures of photons with other quasi-particles.

- Polarons are moving, charged (quasi-) particles that are surrounded by ions in a material.

- Magnons are coherent excitations of electron spins in a material.

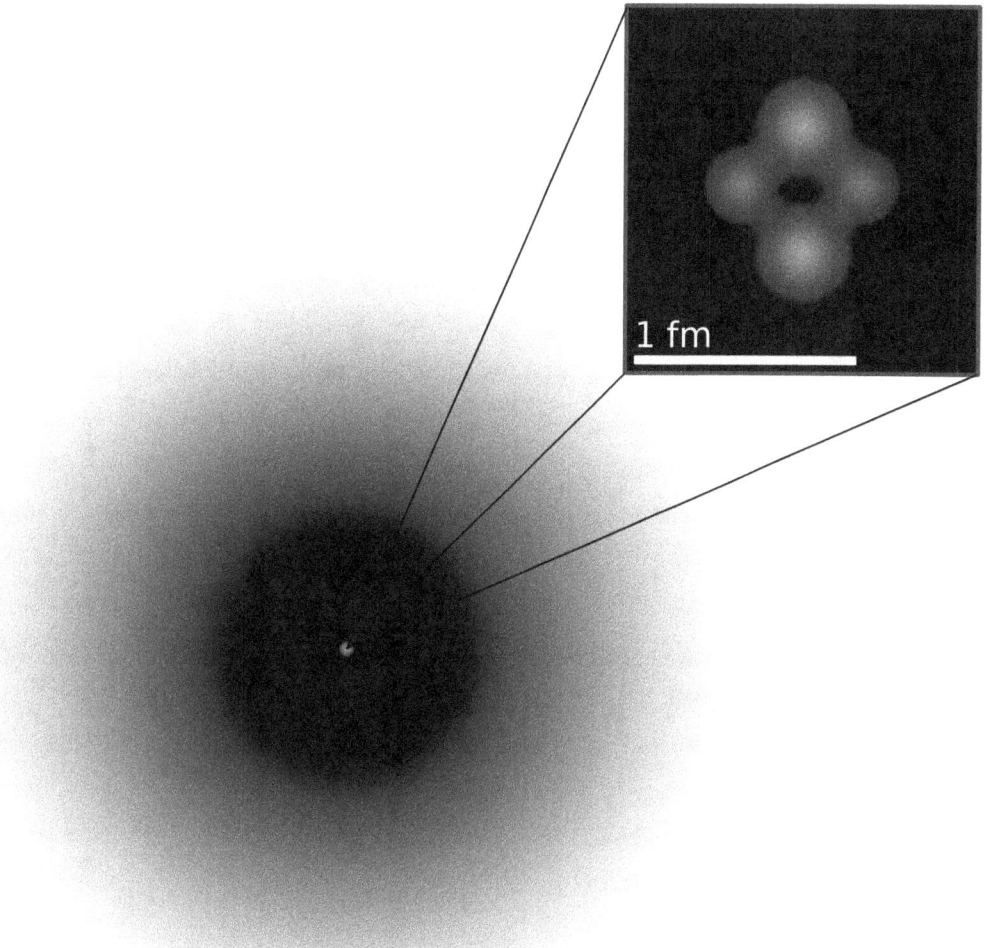

$$1\ \text{Å} = 100{,}000\ \text{fm}$$

A semi-accurate depiction of the helium atom. In the nucleus, the protons are in red and neutrons are in purple. In reality, the nucleus is also spherically symmetrical.

4.4   Other

- An anyon is a generalization of fermion and boson in two-dimensional systems like sheets of graphene that obeys braid statistics.

- A plekton is a theoretical kind of particle discussed as a generalization of the braid statistics of the anyon to dimension > 2.

- A WIMP (weakly interacting massive particle) is any one of a number of particles that might explain dark matter (such as the neutralino or the axion).

- The pomeron, used to explain the elastic scattering of hadrons and the location of Regge poles in Regge theory.

- The skyrmion, a topological solution of the pion field, used to model the low-energy properties of the nucleon, such as the axial vector current coupling and the mass.

- A genon is a particle existing in a closed timelike world line where spacetime is curled as in a Frank Tipler or Ronald Mallett time machine.

- A goldstone boson is a massless excitation of a field that has been spontaneously broken. The pions are quasi-goldstone bosons (quasi- because they are not exactly massless) of the broken chiral isospin symmetry of quantum chromodynamics.

- A goldstino is a goldstone fermion produced by the spontaneous breaking of supersymmetry.

- An instanton is a field configuration which is a local minimum of the Euclidean action. Instantons are used in nonperturbative calculations of tunneling rates.

- A dyon is a hypothetical particle with both electric and magnetic charges.

- A geon is an electromagnetic or gravitational wave which is held together in a confined region by the gravitational attraction of its own field energy.

- An inflaton is the generic name for an unidentified scalar particle responsible for the cosmic inflation.

- A spurion is the name given to a "particle" inserted mathematically into an isospin-violating decay in order to analyze it as though it conserved isospin.

- What is called "true muonium", a bound state of a muon and an antimuon, is a theoretical exotic atom which has never been observed.

4.5 Classification by speed

- A tardyon or bradyon travels slower than light and has a non-zero rest mass.

- A luxon travels at the speed of light and has no rest mass.

- A tachyon (mentioned above) is a hypothetical particle that travels faster than the speed of light and has an imaginary rest mass.

4.6 See also

- Acceleron

- List of baryons

- List of compounds for a list of molecules.

- List of fictional elements, materials, isotopes and atomic particles

- List of mesons

- Periodic table for an overview of atoms.

- Standard Model for the current theory of these particles.

- Table of nuclides

- Timeline of particle discoveries

4.7 References

[1] Observation of a new boson at a mass of 125 GeV with the CMS experiment at the LHC (2013). *arXiv:1207.7235*.

[2] Observation of a new particle in the search for the Standard Model Higgs boson with the ATLAS detector at the LHC (2012). *arXiv:1207.7214*.

[3] B. Kayser, *Two Questions About Neutrinos*, arXiv:1012.4469v1 [hep-ph] (2010).

[4] R. Maartens (2004). *Brane-World Gravity* (PDF). *Living Reviews in Relativity* **7**. p. 7. Also available in web format at http://www.livingreviews.org/lrr-2004-7.

- C. Amsler *et al.* (Particle Data Group) (2008). "Review of Particle Physics". *Physics Letters B* **667** (1–5): 1. Bibcode:2008PhLB..667....1P. doi:10.1016/j.physletb.2008.07.018. *(All information on this list, and more, can be found in the extensive, biannually-updated review by the Particle Data Group)*

Chapter 5

Standard Model

This article is about the Standard Model of particle physics. For other uses, see Standard model (disambiguation).
This article is a non-mathematical general overview of the Standard Model. For a mathematical description, see the article Standard Model (mathematical formulation).
For the Standard Model of Big Bang cosmology, Lambda-CDM model.

The **Standard Model** of particle physics is a theory concerning the electromagnetic, weak, and strong nuclear inter-

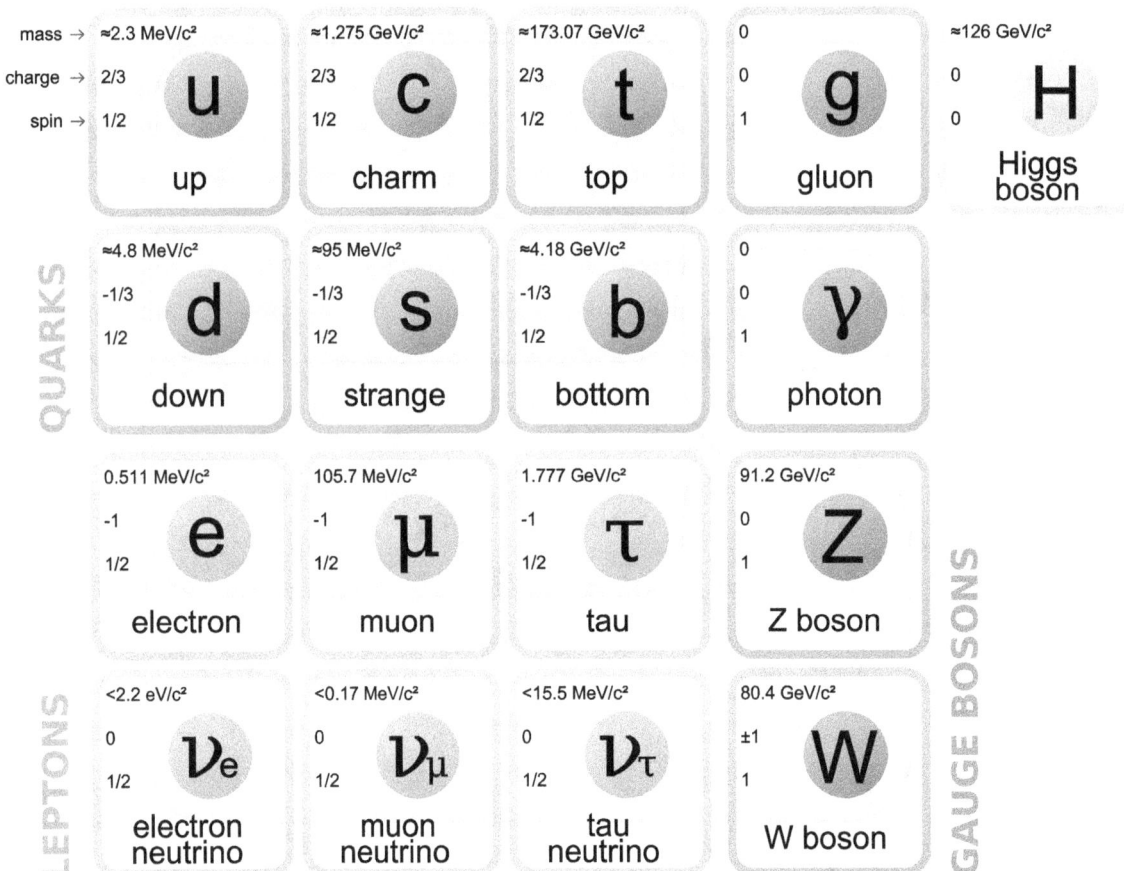

The Standard Model of elementary particles (more schematic depiction), with the three generations of matter, gauge bosons in the fourth column, and the Higgs boson in the fifth.

actions, as well as classifying all the subatomic particles known. It was developed throughout the latter half of the 20th century, as a collaborative effort of scientists around the world.[1] The current formulation was finalized in the mid-1970s upon experimental confirmation of the existence of quarks. Since then, discoveries of the top quark (1995), the tau neutrino (2000), and more recently the Higgs boson (2013), have given further credence to the Standard Model. Because of its success in explaining a wide variety of experimental results, the Standard Model is sometimes regarded as a "theory of almost everything".

Although the Standard Model is believed to be theoretically self-consistent[2] and has demonstrated huge and continued successes in providing experimental predictions, it does leave some phenomena unexplained and it falls short of being a complete theory of fundamental interactions. It does not incorporate the full theory of gravitation[3] as described by general relativity, or account for the accelerating expansion of the universe (as possibly described by dark energy). The model does not contain any viable dark matter particle that possesses all of the required properties deduced from observational cosmology. It also does not incorporate neutrino oscillations (and their non-zero masses).

The development of the Standard Model was driven by theoretical and experimental particle physicists alike. For theorists, the Standard Model is a paradigm of a quantum field theory, which exhibits a wide range of physics including spontaneous symmetry breaking, anomalies, non-perturbative behavior, etc. It is used as a basis for building more exotic models that incorporate hypothetical particles, extra dimensions, and elaborate symmetries (such as supersymmetry) in an attempt to explain experimental results at variance with the Standard Model, such as the existence of dark matter and neutrino oscillations.

5.1 Historical background

The first step towards the Standard Model was Sheldon Glashow's discovery in 1961 of a way to combine the electromagnetic and weak interactions.[4] In 1967 Steven Weinberg[5] and Abdus Salam[6] incorporated the Higgs mechanism[7][8][9] into Glashow's electroweak theory, giving it its modern form.

The Higgs mechanism is believed to give rise to the masses of all the elementary particles in the Standard Model. This includes the masses of the W and Z bosons, and the masses of the fermions, i.e. the quarks and leptons.

After the neutral weak currents caused by Z boson exchange were discovered at CERN in 1973,[10][11][12][13] the electroweak theory became widely accepted and Glashow, Salam, and Weinberg shared the 1979 Nobel Prize in Physics for discovering it. The W and Z bosons were discovered experimentally in 1981, and their masses were found to be as the Standard Model predicted.

The theory of the strong interaction, to which many contributed, acquired its modern form around 1973–74, when experiments confirmed that the hadrons were composed of fractionally charged quarks.

5.2 Overview

At present, matter and energy are best understood in terms of the kinematics and interactions of elementary particles. To date, physics has reduced the laws governing the behavior and interaction of all known forms of matter and energy to a small set of fundamental laws and theories. A major goal of physics is to find the "common ground" that would unite all of these theories into one integrated theory of everything, of which all the other known laws would be special cases, and from which the behavior of all matter and energy could be derived (at least in principle).[14]

5.3 Particle content

The Standard Model includes members of several classes of elementary particles (fermions, gauge bosons, and the Higgs boson), which in turn can be distinguished by other characteristics, such as color charge.

5.3.1 Fermions

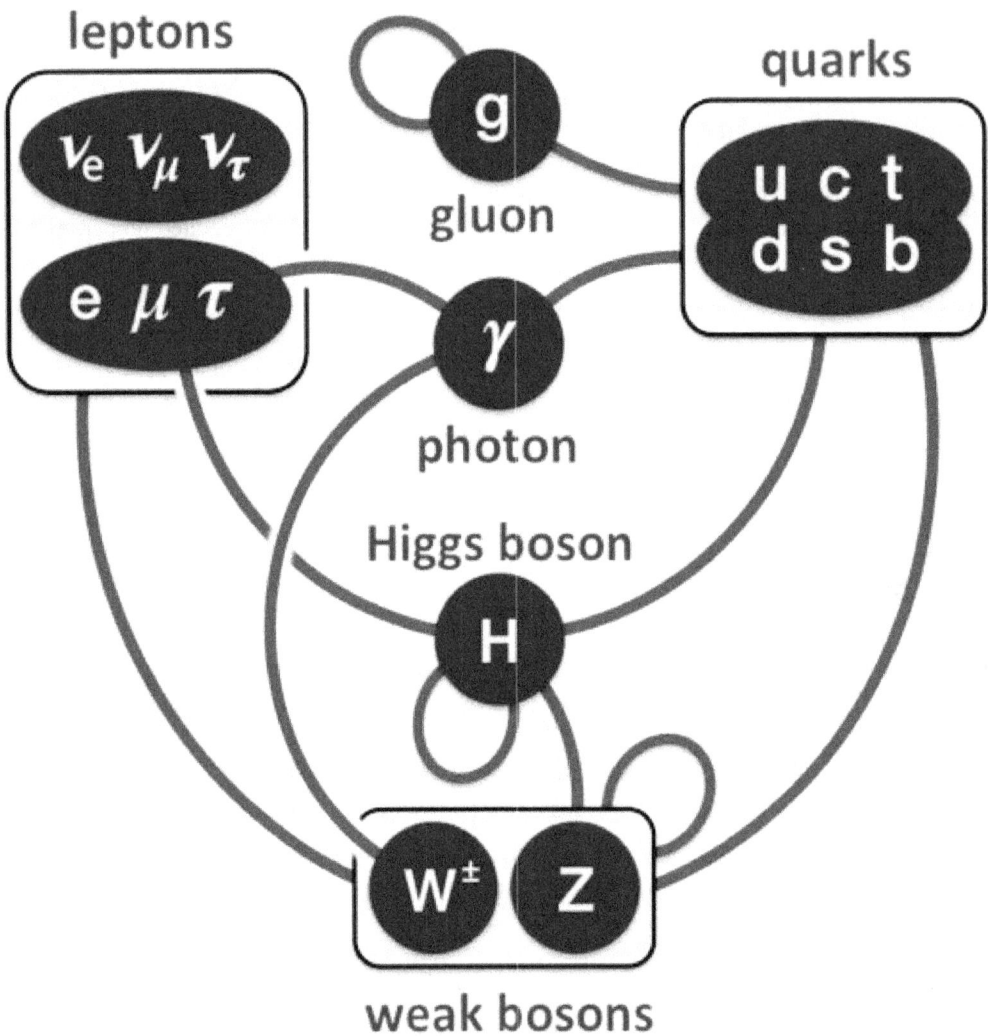

Summary of interactions between particles described by the Standard Model.

The Standard Model includes 12 elementary particles of spin-½ known as fermions. According to the spin-statistics theorem, fermions respect the Pauli exclusion principle. Each fermion has a corresponding antiparticle.

The fermions of the Standard Model are classified according to how they interact (or equivalently, by what charges they carry). There are six quarks (up, down, charm, strange, top, bottom), and six leptons (electron, electron neutrino, muon, muon neutrino, tau, tau neutrino). Pairs from each classification are grouped together to form a generation, with corresponding particles exhibiting similar physical behavior (see table).

The defining property of the quarks is that they carry color charge, and hence, interact via the strong interaction. A phenomenon called color confinement results in quarks being very strongly bound to one another, forming color-neutral composite particles (hadrons) containing either a quark and an antiquark (mesons) or three quarks (baryons). The familiar proton and the neutron are the two baryons having the smallest mass. Quarks also carry electric charge and weak isospin. Hence they interact with other fermions both electromagnetically and via the weak interaction.

The remaining six fermions do not carry colour charge and are called leptons. The three neutrinos do not carry electric charge either, so their motion is directly influenced only by the weak nuclear force, which makes them notoriously difficult to detect. However, by virtue of carrying an electric charge, the electron, muon, and tau all interact electromagnetically.

Each member of a generation has greater mass than the corresponding particles of lower generations. The first generation charged particles do not decay; hence all ordinary (baryonic) matter is made of such particles. Specifically, all atoms consist of electrons orbiting around atomic nuclei, ultimately constituted of up and down quarks. Second and third generations charged particles, on the other hand, decay with very short half lives, and are observed only in very high-energy environments. Neutrinos of all generations also do not decay, and pervade the universe, but rarely interact with baryonic matter.

5.3.2   Gauge bosons

In the Standard Model, gauge bosons are defined as force carriers that mediate the strong, weak, and electromagnetic fundamental interactions.

Interactions in physics are the ways that particles influence other particles. At a macroscopic level, electromagnetism allows particles to interact with one another via electric and magnetic fields, and gravitation allows particles with mass to attract one another in accordance with Einstein's theory of general relativity. The Standard Model explains such forces as resulting from matter particles exchanging other particles, generally referred to as *force mediating particles*. When a force-mediating particle is exchanged, at a macroscopic level the effect is equivalent to a force influencing both of them, and the particle is therefore said to have *mediated* (i.e., been the agent of) that force. The Feynman diagram calculations, which are a graphical representation of the perturbation theory approximation, invoke "force mediating particles", and when applied to analyze high-energy scattering experiments are in reasonable agreement with the data. However, perturbation theory (and with it the concept of a "force-mediating particle") fails in other situations. These include low-energy quantum chromodynamics, bound states, and solitons.

The gauge bosons of the Standard Model all have spin (as do matter particles). The value of the spin is 1, making them bosons. As a result, they do not follow the Pauli exclusion principle that constrains fermions: thus bosons (e.g. photons) do not have a theoretical limit on their spatial density (number per volume). The different types of gauge bosons are described below.

- Photons mediate the electromagnetic force between electrically charged particles. The photon is massless and is well-described by the theory of quantum electrodynamics.

- The W+, W−, and Z gauge bosons mediate the weak interactions between particles of different flavors (all quarks and leptons). They are massive, with the Z being more massive than the W±. The weak interactions involving the W± exclusively act on *left-handed* particles and *right-handed* antiparticles. Furthermore, the W± carries an electric charge of +1 and −1 and couples to the electromagnetic interaction. The electrically neutral Z boson interacts with both left-handed particles and antiparticles. These three gauge bosons along with the photons are grouped together, as collectively mediating the electroweak interaction.

- The eight gluons mediate the strong interactions between color charged particles (the quarks). Gluons are massless. The eightfold multiplicity of gluons is labeled by a combination of color and anticolor charge (e.g. red–antigreen).[nb 1] Because the gluons have an effective color charge, they can also interact among themselves. The gluons and their interactions are described by the theory of quantum chromodynamics.

The interactions between all the particles described by the Standard Model are summarized by the diagrams on the right of this section.

5.3.3   Higgs boson

Main article: Higgs boson

Standard Model Interactions
(Forces Mediated by Gauge Bosons)

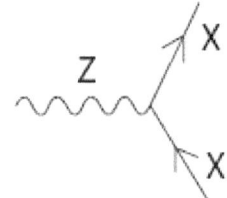

X is any fermion in
the Standard Model.

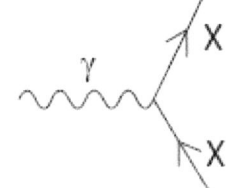

X is electrically charged.

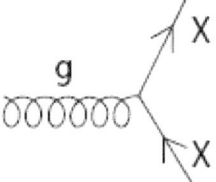

X is any quark.

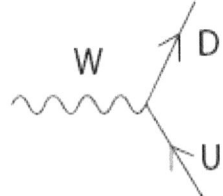

U is a up-type quark;
D is a down-type quark.

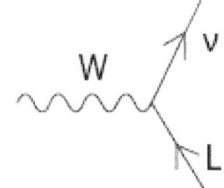

L is a lepton and ν is the
corresponding neutrino.

X is a photon or Z-boson.

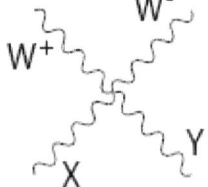

X and Y are any two
electroweak bosons such
that charge is conserved.

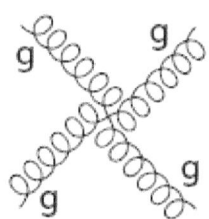

The above interactions form the basis of the standard model. Feynman diagrams in the standard model are built from these vertices. Modifications involving Higgs boson interactions and neutrino oscillations are omitted. The charge of the W bosons is dictated by the fermions they interact with; the conjugate of each listed vertex (i.e. reversing the direction of arrows) is also allowed.

The Higgs particle is a massive scalar elementary particle theorized by Robert Brout, François Englert, Peter Higgs, Gerald Guralnik, C. R. Hagen, and Tom Kibble in 1964 (see 1964 PRL symmetry breaking papers) and is a key building block in the Standard Model.[7][8][9][15] It has no intrinsic spin, and for that reason is classified as a boson (like the gauge bosons, which have integer spin).

The Higgs boson plays a unique role in the Standard Model, by explaining why the other elementary particles, except the photon and gluon, are massive. In particular, the Higgs boson explains why the photon has no mass, while the W and Z bosons are very heavy. Elementary particle masses, and the differences between electromagnetism (mediated by the photon) and the weak force (mediated by the W and Z bosons), are critical to many aspects of the structure of microscopic (and hence macroscopic) matter. In electroweak theory, the Higgs boson generates the masses of the leptons (electron, muon, and tau) and quarks. As the Higgs boson is massive, it must interact with itself.

Because the Higgs boson is a very massive particle and also decays almost immediately when created, only a very high-energy particle accelerator can observe and record it. Experiments to confirm and determine the nature of the Higgs boson using the Large Hadron Collider (LHC) at CERN began in early 2010, and were performed at Fermilab's Tevatron until its closure in late 2011. Mathematical consistency of the Standard Model requires that any mechanism capable of generating the masses of elementary particles become visible at energies above 1.4 TeV;[16] therefore, the LHC (designed to collide two 7 to 8 TeV proton beams) was built to answer the question of whether the Higgs boson actually exists.[17]

On 4 July 2012, the two main experiments at the LHC (ATLAS and CMS) both reported independently that they found a new particle with a mass of about 125 GeV/c^2 (about 133 proton masses, on the order of 10^{-25} kg), which is "consistent with the Higgs boson." Although it has several properties similar to the predicted "simplest" Higgs,[18] they acknowledged that further work would be needed to conclude that it is indeed the Higgs boson, and exactly which version of the Standard Model Higgs is best supported if confirmed.[19][20][21][22][23]

On 14 March 2013 the Higgs Boson was tentatively confirmed to exist.[24]

5.3.4   Total particle count

Counting particles by a rule that distinguishes between particles and their corresponding antiparticles, and among the many color states of quarks and gluons, gives a total of 61 elementary particles.[25]

5.4   Theoretical aspects

Main article: Standard Model (mathematical formulation)

5.4.1   Construction of the Standard Model Lagrangian

Technically, quantum field theory provides the mathematical framework for the Standard Model, in which a Lagrangian controls the dynamics and kinematics of the theory. Each kind of particle is described in terms of a dynamical field that pervades space-time. The construction of the Standard Model proceeds following the modern method of constructing most field theories: by first postulating a set of symmetries of the system, and then by writing down the most general renormalizable Lagrangian from its particle (field) content that observes these symmetries.

The global Poincaré symmetry is postulated for all relativistic quantum field theories. It consists of the familiar translational symmetry, rotational symmetry and the inertial reference frame invariance central to the theory of special relativity. The local SU(3)×SU(2)×U(1) gauge symmetry is an internal symmetry that essentially defines the Standard Model. Roughly, the three factors of the gauge symmetry give rise to the three fundamental interactions. The fields fall into different representations of the various symmetry groups of the Standard Model (see table). Upon writing the most general Lagrangian, one finds that the dynamics depend on 19 parameters, whose numerical values are established by experiment. The parameters are summarized in the table above (note: with the Higgs mass is at 125 GeV, the Higgs self-coupling strength $\lambda \sim 1/8$).

Quantum chromodynamics sector

Main article: Quantum chromodynamics

The quantum chromodynamics (QCD) sector defines the interactions between quarks and gluons, with SU(3) symmetry, generated by T^a. Since leptons do not interact with gluons, they are not affected by this sector. The Dirac Lagrangian of the quarks coupled to the gluon fields is given by

$$\mathcal{L}_{QCD} = i\overline{U}(\partial_\mu - ig_s G_\mu^a T^a)\gamma^\mu U + i\overline{D}(\partial_\mu - ig_s G_\mu^a T^a)\gamma^\mu D.$$

G_μ^a is the SU(3) gauge field containing the gluons, γ^μ are the Dirac matrices, D and U are the Dirac spinors associated with up- and down-type quarks, and g_s is the strong coupling constant.

Electroweak sector

Main article: Electroweak interaction

The electroweak sector is a Yang–Mills gauge theory with the simple symmetry group U(1)×SU(2)L,

$$\mathcal{L}_{EW} = \sum_\psi \bar{\psi} \gamma^\mu \left(i\partial_\mu - g' \frac{1}{2} Y_W B_\mu - g \frac{1}{2} \vec{\tau}_L \vec{W}_\mu \right) \psi$$

where $B\mu$ is the U(1) gauge field; YW is the weak hypercharge—the generator of the U(1) group; $\vec{W}_\mu$ is the three-component SU(2) gauge field; $\vec{\tau}_L$ are the Pauli matrices—infinitesimal generators of the SU(2) group. The subscript L indicates that they only act on left fermions; g' and g are coupling constants.

Higgs sector

Main article: Higgs mechanism

In the Standard Model, the Higgs field is a complex scalar of the group SU(2)L:

$$\varphi = \frac{1}{\sqrt{2}} \begin{pmatrix} \varphi^+ \\ \varphi^0 \end{pmatrix},$$

where the indices + and 0 indicate the electric charge (Q) of the components. The weak isospin (YW) of both components is 1.

Before symmetry breaking, the Higgs Lagrangian is:

$$\mathcal{L}_H = \varphi^\dagger \left(\partial^\mu - \frac{i}{2} \left(g' Y_W B^\mu + g \vec{\tau} \vec{W}^\mu \right) \right) \left(\partial_\mu + \frac{i}{2} \left(g' Y_W B_\mu + g \vec{\tau} \vec{W}_\mu \right) \right) \varphi - \frac{\lambda^2}{4} \left(\varphi^\dagger \varphi - v^2 \right)^2,$$

which can also be written as:

$$\mathcal{L}_H = \left| \left(\partial_\mu + \frac{i}{2} \left(g' Y_W B_\mu + g \vec{\tau} \vec{W}_\mu \right) \right) \varphi \right|^2 - \frac{\lambda^2}{4} \left(\varphi^\dagger \varphi - v^2 \right)^2.$$

5.5 Fundamental forces

Main article: Fundamental interaction

The Standard Model classified all four fundamental forces in nature. In the Standard Model, a force is described as an exchange of bosons between the objects affected, such as a photon for the electromagnetic force and a gluon for the strong interaction. Those particles are called force carriers.[26]

5.6   Tests and predictions

The Standard Model (SM) predicted the existence of the W and Z bosons, gluon, and the top and charm quarks before these particles were observed. Their predicted properties were experimentally confirmed with good precision. To give an idea of the success of the SM, the following table compares the measured masses of the W and Z bosons with the masses predicted by the SM:

The SM also makes several predictions about the decay of Z bosons, which have been experimentally confirmed by the Large Electron-Positron Collider at CERN.

In May 2012 BaBar Collaboration reported that their recently analyzed data may suggest possible flaws in the Standard Model of particle physics.[28][29] These data show that a particular type of particle decay called "B to D-star-tau-nu" happens more often than the Standard Model says it should. In this type of decay, a particle called the B-bar meson decays into a D meson, an antineutrino and a tau-lepton. While the level of certainty of the excess (3.4 sigma) is not enough to claim a break from the Standard Model, the results are a potential sign of something amiss and are likely to impact existing theories, including those attempting to deduce the properties of Higgs bosons.[30]

On December 13, 2012, physicists reported the constancy, over space and time, of a basic physical constant of nature that supports the *standard model of physics*. The scientists, studying methanol molecules in a distant galaxy, found the change $(\Delta\mu/\mu)$ in the proton-to-electron mass ratio μ to be equal to "$(0.0 \pm 1.0) \times 10^{-7}$ at redshift z = 0.89" and consistent with "a null result".[31][32]

5.7   Challenges

See also: Physics beyond the Standard Model

Self-consistency of the Standard Model (currently formulated as a non-abelian gauge theory quantized through path-integrals) has not been mathematically proven. While regularized versions useful for approximate computations (for example lattice gauge theory) exist, it is not known whether they converge (in the sense of S-matrix elements) in the limit that the regulator is removed. A key question related to the consistency is the Yang–Mills existence and mass gap problem.

Experiments indicate that neutrinos have mass, which the classic Standard Model did not allow.[33] To accommodate this finding, the classic Standard Model can be modified to include neutrino mass.

If one insists on using only Standard Model particles, this can be achieved by adding a non-renormalizable interaction of leptons with the Higgs boson.[34] On a fundamental level, such an interaction emerges in the seesaw mechanism where heavy right-handed neutrinos are added to the theory. This is natural in the left-right symmetric extension of the Standard Model[35][36] and in certain grand unified theories.[37] As long as new physics appears below or around 10^{14} GeV, the neutrino masses can be of the right order of magnitude.

Theoretical and experimental research has attempted to extend the Standard Model into a Unified field theory or a Theory of everything, a complete theory explaining all physical phenomena including constants. Inadequacies of the Standard Model that motivate such research include:

- It does not attempt to explain gravitation, although a theoretical particle known as a graviton would help explain it, and unlike for the strong and electroweak interactions of the Standard Model, there is no known way of describing general relativity, the canonical theory of gravitation, consistently in terms of quantum field theory. The reason for this is, among other things, that quantum field theories of gravity generally break down before reaching the Planck scale. As a consequence, we have no reliable theory for the very early universe;

- Some consider it to be *ad hoc* and inelegant, requiring 19 numerical constants whose values are unrelated and arbitrary. Although the Standard Model, as it now stands, can explain why neutrinos have masses, the specifics of neutrino mass are still unclear. It is believed that explaining neutrino mass will require an additional 7 or 8 constants, which are also arbitrary parameters;

- The Higgs mechanism gives rise to the hierarchy problem if some new physics (coupled to the Higgs) is present at high energy scales. In these cases in order for the weak scale to be much smaller than the Planck scale, severe fine tuning of the parameters is required; there are, however, other scenarios that include quantum gravity in which such fine tuning can be avoided.[38]There are also issues of Quantum triviality, which suggests that it may not be possible to create a consistent quantum field theory involving elementary scalar particles.

- It should be modified so as to be consistent with the emerging "Standard Model of cosmology." In particular, the Standard Model cannot explain the observed amount of cold dark matter (CDM) and gives contributions to dark energy which are many orders of magnitude too large. It is also difficult to accommodate the observed predominance of matter over antimatter (matter/antimatter asymmetry). The isotropy and homogeneity of the visible universe over large distances seems to require a mechanism like cosmic inflation, which would also constitute an extension of the Standard Model.

- The existence of ultra-high-energy cosmic rays are difficult to explain under the Standard Model.

Currently, no proposed Theory of Everything has been widely accepted or verified.

5.8 See also

- Fundamental interaction:

  - Quantum electrodynamics

  - Strong interaction: Color charge, Quantum chromodynamics, Quark model

  - Weak interaction: Electroweak theory, Fermi theory of beta decay, Weak hypercharge, Weak isospin

- Gauge theory: Nontechnical introduction to gauge theory

- Generation

- Higgs mechanism: Higgs boson, Higgsless model

- J. C. Ward

- J. J. Sakurai Prize for Theoretical Particle Physics

- Lagrangian

- Open questions: BTeV experiment, CP violation, Neutrino masses, Quark matter, Quantum triviality

- Penguin diagram

- Quantum field theory

- Standard Model: Mathematical formulation of, Physics beyond the Standard Model

5.9 Notes and references

[1] Technically, there are nine such color–anticolor combinations. However, there is one color-symmetric combination that can be constructed out of a linear superposition of the nine combinations, reducing the count to eight.

5.10   References

[1] R. Oerter (2006). *The Theory of Almost Everything: The Standard Model, the Unsung Triumph of Modern Physics* (Kindle ed.). Penguin Group. p. 2. ISBN 0-13-236678-9.

[2] In fact, there are mathematical issues regarding quantum field theories still under debate (see e.g. Landau pole), but the predictions extracted from the Standard Model by current methods applicable to current experiments are all self-consistent. For a further discussion see e.g. Chapter 25 of R. Mann (2010). *An Introduction to Particle Physics and the Standard Model*. CRC Press. ISBN 978-1-4200-8298-2.

[3] Sean Carroll, Ph.D., Cal Tech, 2007, The Teaching Company, *Dark Matter, Dark Energy: The Dark Side of the Universe*, Guidebook Part 2 page 59, Accessed Oct. 7, 2013, "...Standard Model of Particle Physics: The modern theory of elementary particles and their interactions ... It does not, strictly speaking, include gravity, although it's often convenient to include gravitons among the known particles of nature..."

[4] S.L. Glashow (1961). "Partial-symmetries of weak interactions". *Nuclear Physics* **22** (4): 579–588. Bibcode:1961NucPh..22..579G. doi:10.1016/0029-5582(61)90469-2.

[5] S. Weinberg (1967). "A Model of Leptons". *Physical Review Letters* **19** (21): 1264–1266. Bibcode:1967PhRvL..19.1264W. doi:10.1103/PhysRevLett.19.1264.

[6] A. Salam (1968). N. Svartholm, ed. *Elementary Particle Physics: Relativistic Groups and Analyticity*. Eighth Nobel Symposium. Stockholm: Almquvist and Wiksell. p. 367.

[7] F. Englert, R. Brout (1964). "Broken Symmetry and the Mass of Gauge Vector Mesons". *Physical Review Letters* **13** (9): 321–323. Bibcode:1964PhRvL..13..321E. doi:10.1103/PhysRevLett.13.321.

[8] P.W. Higgs (1964). "Broken Symmetries and the Masses of Gauge Bosons". *Physical Review Letters* **13** (16): 508–509. Bibcode:1964PhRvL..13..508H. doi:10.1103/PhysRevLett.13.508.

[9] G.S. Guralnik, C.R. Hagen, T.W.B. Kibble (1964). "Global Conservation Laws and Massless Particles". *Physical Review Letters* **13** (20): 585–587. Bibcode:1964PhRvL..13..585G. doi:10.1103/PhysRevLett.13.585.

[10] F.J. Hasert et al. (1973). "Search for elastic muon-neutrino electron scattering". *Physics Letters B* **46**(1):121. Bibcode:1973PhLB... doi:10.1016/0370-2693(73)90494-2.

[11] F.J. Hasert et al. (1973). "Observation of neutrino-like interactions without muon or electron in the Gargamelle neutrino experiment". *Physics Letters B* **46** (1): 138. Bibcode:1973PhLB...46..138H. doi:10.1016/0370-2693(73)90499-1.

[12] F.J. Hasert et al. (1974). "Observation of neutrino-like interactions without muon or electron in the Gargamelle neutrino experiment". *Nuclear Physics B* **73** (1): 1. Bibcode:1974NuPhB..73....1H. doi:10.1016/0550-3213(74)90038-8.

[13] D. Haidt (4 October 2004). "The discovery of the weak neutral currents". *CERN Courier*. Retrieved 8 May 2008.

[14] "Details can be worked out if the situation is simple enough for us to make an approximation, which is almost never, but often we can understand more or less what is happening." from *The Feynman Lectures on Physics*, Vol 1. pp. 2–7

[15] G.S. Guralnik (2009). "The History of the Guralnik, Hagen and Kibble development of the Theory of Spontaneous Symmetry Breaking and Gauge Particles". *International Journal of Modern Physics A* **24** (14): 2601–2627. arXiv:0907.3466. Bibcode:2009IJMPA..24.2601G. doi:10.1142/S0217751X09045431.

[16] B.W. Lee, C. Quigg, H.B. Thacker (1977). "Weak interactions at very high energies: The role of the Higgs-boson mass". *Physical Review D* **16** (5): 1519–1531. Bibcode:1977PhRvD..16.1519L. doi:10.1103/PhysRevD.16.1519.

[17] "Huge $10 billion collider resumes hunt for 'God particle'". CNN. 11 November 2009. Retrieved 2010-05-04.

[18] M. Strassler (10 July 2012). "Higgs Discovery: Is it a Higgs?". Retrieved 2013-08-06.

[19] "CERN experiments observe particle consistent with long-sought Higgs boson". CERN. 4 July 2012. Retrieved 2012-07-04.

[20] "Observation of a New Particle with a Mass of 125 GeV". CERN. 4 July 2012. Retrieved 2012-07-05.

[21] "ATLAS Experiment". ATLAS. 1 January 2006. Retrieved 2012-07-05.

[22] "Confirmed: CERN discovers new particle likely to be the Higgs boson". *YouTube*. Russia Today. 4 July 2012. Retrieved 2013-08-06.

[23] D. Overbye (4 July 2012). "A New Particle Could Be Physics' Holy Grail". *New York Times*. Retrieved 2012-07-04.

[24] "New results indicate that new particle is a Higgs boson". CERN. 14 March 2013. Retrieved 2013-08-06.

[25] S. Braibant, G. Giacomelli, M. Spurio (2009). *Particles and Fundamental Interactions: An Introduction to Particle Physics*. Springer. pp. 313–314. ISBN 978-94-007-2463-1.

[26] http://home.web.cern.ch/about/physics/standard-model Official CERN website

[27] http://www.pha.jhu.edu/~{}dfehling/particle.gif

[28] "BABAR Data in Tension with the Standard Model". SLAC. 31 May 2012. Retrieved 2013-08-06.

[29] BaBar Collaboration (2012). "Evidence for an excess of $B \to D^{(*)} \tau^- \nu\tau$ decays". *Physical Review Letters* **109** (10): 101802. arXiv:1205.5442. Bibcode:2012PhRvL.109j1802L. doi:10.1103/PhysRevLett.109.101802.

[30] "BaBar data hint at cracks in the Standard Model". *e! Science News*. 18 June 2012. Retrieved 2013-08-06.

[31] J. Bagdonaite et al. (2012). "A Stringent Limit on a Drifting Proton-to-Electron Mass Ratio from Alcohol in the Early Universe". *Science* **339** (6115): 46. Bibcode:2013Sci...339...46B. doi:10.1126/science.1224898.

[32] C. Moskowitz (13 December 2012). "Phew! Universe's Constant Has Stayed Constant". Space.com. Retrieved 2012-12-14.

[33] "Particle chameleon caught in the act of changing". CERN. 31 May 2010. Retrieved 2012-07-05.

[34] S.Weinberg(1979). "Baryon and Lepton Nonconserving Processes".*Physical Review Letters***43**(21):1566.Bibcode:1979PhRvW. doi:10.1103/PhysRevLett.43.1566.

[35] P.Minkowski(7..421M.

doi:10.1016/0370-2693(77)90435-X.

[36] R. N. Mohapatra, G. Senjanovic (1980). "Neutrino Mass and Spontaneous Parity Nonconservation". *Physical Review Letters* **44** (14): 912–915. Bibcode:1980PhRvL..44..912M. doi:10.1103/PhysRevLett.44.912.

[37] M. Gell-Mann, P. Ramond and R. Slansky (1979). F. van Nieuwenhuizen and D. Z. Freedman, ed. *Supergravity*. North Holland. pp. 315–321. ISBN 0-444-85438-X.

[38] Salvio,Strumia(2014-03-17)."Agravity".*JHEP1406(2014)080*.arXiv:1403.4226.Bibcode:2014JHEP...06..080S.doi:10.)080.

5.11 Further reading

- R. Oerter (2006). *The Theory of Almost Everything: The Standard Model, the Unsung Triumph of Modern Physics*. Plume.

- B.A. Schumm (2004). *Deep Down Things: The Breathtaking Beauty of Particle Physics*. Johns Hopkins University Press. ISBN 0-8018-7971-X.

- "The Standard Model of Particle Physics Interactive Graphic".

Introductory textbooks

- I. Aitchison, A. Hey (2003). *Gauge Theories in Particle Physics: A Practical Introduction*. Institute of Physics. ISBN 978-0-585-44550-2.

- W. Greiner, B. Müller (2000). *Gauge Theory of Weak Interactions*. Springer. ISBN 3-540-67672-4.

- G.D. Coughlan, J.E. Dodd, B.M. Gripaios (2006). *The Ideas of Particle Physics: An Introduction for Scientists*. Cambridge University Press.

- D.J. Griffiths (1987). *Introduction to Elementary Particles*. John Wiley & Sons. ISBN 0-471-60386-4.

- G.L. Kane (1987). *Modern Elementary Particle Physics*. Perseus Books. ISBN 0-201-11749-5.

Advanced textbooks

- T.P. Cheng, L.F. Li (2006). *Gauge theory of elementary particle physics*. Oxford University Press. ISBN 0-19-851961-3. Highlights the gauge theory aspects of the Standard Model.

- J.F. Donoghue, E. Golowich, B.R. Holstein (1994). *Dynamics of the Standard Model*. Cambridge University Press. ISBN 978-0-521-47652-2. Highlights dynamical and phenomenological aspects of the Standard Model.

- L. O'Raifeartaigh (1988). *Group structure of gauge theories*. Cambridge University Press. ISBN 0-521-34785-8.

- Nagashima Y. Elementary Particle Physics: Foundations of the Standard Model, Volume 2. (Wiley 2013) 920 panubi

- Schwartz, M.D. Quantum Field Theory and the Standard Model (Cambridge University Press 2013) 952 pages

- Langacker P. The standard model and beyond. (CRC Press, 2010) 670 pages Highlights group-theoretical aspects of the Standard Model.

Journal articles

- E.S.Abers,B.W.Lee(1973). "Gauge theories".*Physics Reports***9**:1–141.Bibcode:1973PhR.....9....1A.doi:10.10 1573(73)90027-6.

- M. Baak et al. (2012). "The Electroweak Fit of the Standard Model after the Discovery of a New Boson at the LHC". *The European Physical Journal C* **72** (11). arXiv:1209.2716. Bibcode:2012EPJC...72.2205B. doi:10.1140/epjc/s10052-012-2205-9.

- Y. Hayato et al. (1999). "Search for Proton Decay through $p \rightarrow \nu K^+$ in a Large Water Cherenkov Detector". *Physical Review Letters***83**(8):1529.arXiv:hep-ex/9904020.Bibcode:1999PhRvL..83.1529H.doi:10.1103/Phys.

- S.F. Novaes (2000). "Standard Model: An Introduction". arXiv:hep-ph/0001283 [hep-ph].

- D.P. Roy (1999). "Basic Constituents of Matter and their Interactions — A Progress Report". arXiv:hep-ph/9912523 [hep-ph].

- F. Wilczek (2004). "The Universe Is A Strange Place". *Nuclear Physics B - Proceedings Supplements* **134**: 3. arXiv:astro-ph/0401347. Bibcode:2004NuPhS.134....3W. doi:10.1016/j.nuclphysbps.2004.08.001.

5.12 External links

- "The Standard Model explained in Detail by CERN's John Ellis" omega tau podcast.

- "LHC sees hint of lightweight Higgs boson" "New Scientist".

- "Standard Model may be found incomplete," *New Scientist*.

- "Observation of the Top Quark" at Fermilab.

- "The Standard Model Lagrangian." After electroweak symmetry breaking, with no explicit Higgs boson.

- "Standard Model Lagrangian" with explicit Higgs terms. PDF, PostScript, and LaTeX versions.

- "The particle adventure." Web tutorial.

- Nobes, Matthew (2002) "Introduction to the Standard Model of Particle Physics" on Kuro5hin: Part 1, Part 2, Part 3a, Part 3b.

- "The Standard Model" The Standard Model on the CERN web site explains how the basic building blocks of matter interact, governed by four fundamental forces.

Chapter 6

Charge radius

The **rms charge radius** is a measure of the size of an atomic nucleus, particularly of a proton or a deuteron. It can be measured by the scattering of electrons by the nucleus and also inferred from the effects of finite nuclear size on electron energy levels as measured in atomic spectra.

6.1 Definition

The problem of defining a radius for the atomic nucleus is similar to the problem of atomic radius, in that neither atoms nor their nuclei have definite boundaries. However, the nucleus can be modeled as a sphere of positive charge for the interpretation of electron scattering experiments: because there is no definite boundary to the nucleus, the electrons "see" a range of cross-sections, for which a mean can be taken. The qualification of "rms" (for "root mean square") arises because it is the nuclear cross-section, proportional to the square of the radius, which is determining for electron scattering.

This definition of charge radius can also be applied to composite hadrons such as a proton, neutron, pion, or kaon, that are made up of more than one quark. In the case of an anti-matter baryon (e.g. an anti-proton), and some particles with a net zero electric charge, the composite particle must be modeled as a sphere of negative rather than positive electric charge for the interpretation of electron scattering experiments. In these cases, the square of the charge radius of the particle is defined to be negative, with the same absolute value with units of length squared equal to the positive squared charge radius that it would have had if it was identical in all other respects but each quark in the particle had the opposite electric charge (with the charge radius itself having a value that is an imaginary number with units of length).[1] It is customary when charge radius takes an imaginary numbered value to report the negative valued square of the charge radius, rather than the charge radius itself, for a particle.

The best known particle with a negative squared charge radius is the neutron. The heuristic explanation for why the squared charge radius of a neutron is negative, despite its overall neutral electric charge, is that this is the case because its negatively charged down quarks are, on average, located in the outer part of the neutron, while its positively charged up quark is, on average, located towards the center of the neutron. This asymmetric distribution of charge within the particle gives rise to a small negative squared charge radius for the particle as a whole. But, this is only the simplest of a variety of theoretical models, some of which are more elaborate, that are used to explain this property of a neutron.[2]

For deuterons and higher nuclei, it is conventional to distinguish between the scattering charge radius, r_d (obtained from scattering data), and the bound-state charge radius, R_d, which includes the Darwin–Foldy term to account for the behaviour of the anomalous magnetic moment in an electromagnetic field[3][4] and which is appropriate for treating spectroscopic data.[5] The two radii are related by

$$R_d = \sqrt{r_d^2 + \frac{3}{4}\left(\frac{m_e}{m_d}\right)^2\left(\frac{\lambda_C}{2\pi}\right)^2}$$

where m_e and m_d are the masses of the electron and the deuteron respectively while λC is the Compton wavelength of the electron.[5] For the proton, the two radii are the same.[5]

6.2 History

Main article: Geiger–Marsden experiment

The first estimate of a nuclear charge radius was made by Hans Geiger and Ernest Marsden in 1909,[6] under the direction of Ernest Rutherford at the Physical Laboratories of the University of Manchester, UK. The famous experiment involved the scattering of α-particles by gold foil, with some of the particles being scattered through angles of more than 90°, that is coming back to the same side of the foil as the α-source. Rutherford was able to put an upper limit on the radius of the gold nucleus of 34 femtometres.[7]

Later studies found an empirical relation between the charge radius and the mass number, A, for heavier nuclei ($A > 20$):

$$R \approx r_0 A^{1/3}$$

where r_0 is an empirical constant of 1.2–1.5 fm. This gives a charge radius for the gold nucleus ($A = 197$) of about 7.5 fm.[8]

6.3 Modern measurements

Modern direct measurements are based on the scattering of electrons by nuclei.[9][10] There is most interest in knowing the charge radii of protons and deuterons, as these can be compared with the spectrum of atomic hydrogen/deuterium: the nonzero size of the nucleus causes a shift in the electronic energy levels which shows up as a change in the frequency of the spectral lines.[5] Such comparisons are a test of quantum electrodynamics (QED). Since 2002, the proton and deuteron charge radii have been independently refined parameters in the CODATA set of recommended values for physical constants, that is both scattering data and spectroscopic data are used to determine the recommended values.[11]

The 2010 CODATA recommended values are:

proton: $R_p = 0.8775(51)*10^{-15}$ m
deuteron: $R_d = 2.1424(21)*10^{-15}$ m

Recent work on the spectrum of muonic hydrogen (an exotic atom consisting of a proton and a negative muon) indicates a significantly lower value for the proton charge radius, 0.84087(39) fm: the reason for this discrepancy is not clear.[12]

6.4 References

[1] See, e.g., Abouzaid, et al., "A Measurement of the K0 Charge Radius and a CP Violating Asymmetry Together with a Search for CP Violating E1 Direct Photon Emission in the Rare Decay KL->pi+pi-e+e-", Phys.Rev.Lett.96:101801 (2006) DOI: 10.1103/PhysRevLett.96.101801 http://arxiv.org/abs/hep-ex/0508010 (determining that the neutral kaon has a negative mean squared charge radius of -0.077 ± 0.007(stat) ± 0.011(syst)fm^2).

[2] See, e.g., J. Byrne, "The mean square charge radius of the neutron", Neutron News Vol. 5, Issue 4, pg. 15-17 (1994) (comparing different theoretical explanations for the neutron's observed negative squared charge radius to the data) DOI: 10.1080/10448639408217664 http://www.tandfonline.com/doi/abs/10.1080/10448639408217664#.U3GYaPldVUA

[3] Foldy, L.L. (1958), "Neutron–Electron Interaction", *Rev. Mod. Phys.* **30**:471–81, Bibcode:1958RvMP...30..471F, doi:10.1.471.

[4] Friar, J. L.; Martorell, J.; Sprung, D. W. L. (1997), "Nuclear sizes and the isotope shift", *Phys. Rev. A* **56**: 4579–86, arXiv:nucl-th/9707016, Bibcode:1997PhRvA..56.4579F, doi:10.1103/PhysRevA.56.4579.

[5] Mohr, Peter J.; Taylor, Barry N. (1999). "CODATA recommended values of the fundamental physical constants: 1998". *J. Phys. Chem. Ref. Data* **28** (6): 1713–1852. doi:10.1103/RevModPhys.72.351.

[6] Geiger, H.; Marsden, E. (1909), "On a Diffuse Reflection of the α-Particles", *Proceedings of the Royal Society A* **82**: 495–500, Bibcode:1909RSPSA..82..495G, doi:10.1098/rspa.1909.0054.

[7] Rutherford, E. (1911), "The Scattering of α and β Particles by Matter and the Structure of the Atom", *Phil. Mag., Ser. 6* **21**: 669–88, doi:10.1080/14786440508637080.

[8] Blatt, John M.; Weisskopf, Victor F. (1952), *Theoretical Nuclear Physics*, New York: Wiley, pp. 14–16.

[9]Sick,Ingo(2003), "On the rms-radius of the proton",*Phys.Lett.B***576**(1–2):62–67,arXiv:nucl-ex/0310008,Bibcode:2003PhS, doi:10.1016/j.physletb.2003.09.092.

[10]Sick,Ingo;Trautmann,Dirk(1998), "On the rms radius of the deuteron",*Nucl.Phys.A***637**(4):559–75,Bibcode:1998NuPh59S, doi:10.1016/S0375-9474(98)00334-0.

[11] Mohr, Peter J.; Taylor, Barry N. (2005). "CODATA recommended values of the fundamental physical constants: 2002". *Rev. Mod. Phys.* **77** (1): 1–107. Bibcode:2005RvMP...77....1M. doi:10.1103/RevModPhys.77.1.

[12] Antognini, A.; Nez, F.; Schuhmann, K.; Amaro, F. D.; Biraben, F.; Cardoso, J. M. R.; Covita, D. S.; Dax, A.; Dhawan, S.; Diepold, M.; Fernandes, L. M. P.; Giesen, A.; Gouvea, A. L.; Graf, T.; Hänsch, T. W.; Indelicato, P.; Julien, L.; Kao, C. -Y.; Knowles, P.; Kottmann, F.; Le Bigot, E. -O.; Liu, Y. -W.; Lopes, J. A. M.; Ludhova, L.; Monteiro, C. M. B.; Mulhauser, F.; Nebel, T.; Rabinowitz, P.; Dos Santos, J. M. F.; Schaller, L. A. (2013). "Proton Structure from the Measurement of 2S-2P Transition Frequencies of Muonic Hydrogen". *Science* **339** (6118): 417–420. doi:10.1126/science.1230016. PMID 23349284.

Chapter 7

Proton decay

This article is about decay of protons into subatomic particles. For the type of radioactive decay in which a nucleus ejects a proton, see Proton emission.

In particle physics, **proton decay** is a hypothetical form of radioactive decay in which the proton decays into lighter subatomic particles, such as a neutral pion and a positron.[1] There is currently no experimental evidence that proton decay occurs.

In the Standard Model, protons, a type of baryon, are theoretically stable because baryon number (quark number) is conserved (under normal circumstances; however, see chiral anomaly). Therefore, protons will not decay into other particles on their own, because they are the lightest (and therefore least energetic) baryon.

Some beyond-the-Standard Model grand unified theories (GUTs) explicitly break the baryon number symmetry, allowing protons to decay via the Higgs particle, magnetic monopoles or new X bosons. Proton decay is one of the few unobserved effects of the various proposed GUTs. To date, all attempts to observe these events have failed.

7.1 Baryogenesis

Main article: Baryogenesis

One of the outstanding problems in modern physics is the predominance of matter over antimatter in the universe. The universe, as a whole, seems to have a nonzero positive baryon number density — that is, matter exists. Since it is assumed in cosmology that the particles we see were created using the same physics we measure today, it would normally be expected that the overall baryon number should be zero, as matter and antimatter should have been created in equal amounts. This has led to a number of proposed mechanisms for symmetry breaking that favour the creation of normal matter (as opposed to antimatter) under certain conditions. This imbalance would have been exceptionally small, on the order of 1 in every 10000000000 (10^{10}) particles a small fraction of a second after the Big Bang, but after most of the matter and antimatter annihilated, what was left over was all the baryonic matter in the current universe, along with a much greater number of bosons. Experiments reported in 2010 at Fermilab, however, seem to show that this imbalance is much greater than previously assumed. In an experiment involving a series of particle collisions, the amount of generated matter was approximately 1% larger than the amount of generated antimatter. The reason for this discrepancy is yet unknown.[2]

Most grand unified theories explicitly break the baryon number symmetry, which would account for this discrepancy, typically invoking reactions mediated by very massive X bosons (X) or massive Higgs bosons (H0). The rate at which these events occur is governed largely by the mass of the intermediate X or H0 particles, so by assuming these reactions are responsible for the majority of the baryon number seen today, a maximum mass can be calculated above which the rate would be too slow to explain the presence of matter today. These estimates predict that a large volume of material will occasionally exhibit a spontaneous proton decay.

51

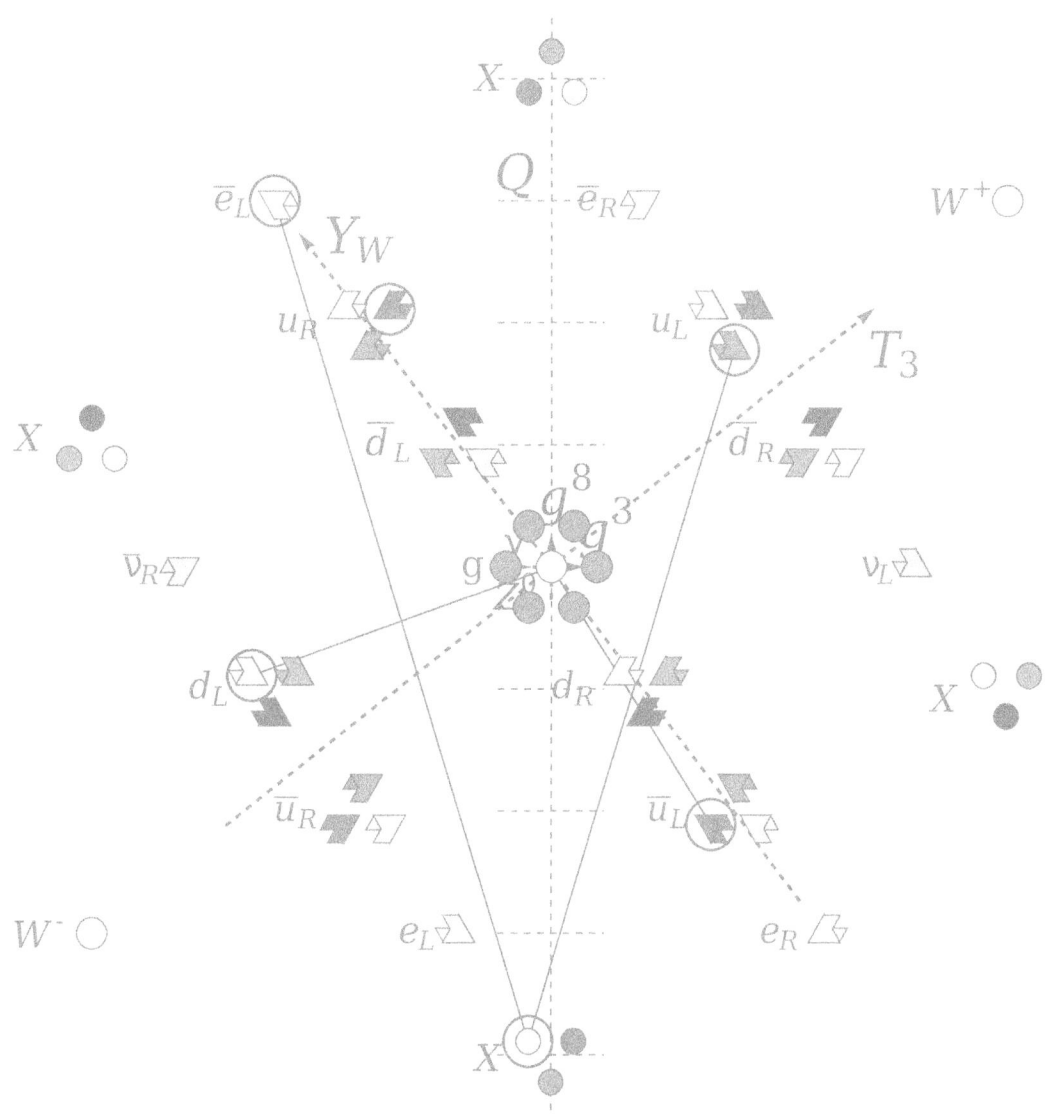

The pattern of weak isospins, weak hypercharges, and color charges for particles in the Georgi–Glashow model. Here, a proton, consisting of two up quarks and a down, decays into a pi meson, consisting of an up and anti-up, and a positron, via an X boson with electric charge −4/3.

7.2   Experimental evidence

Proton decay is one of the few unobserved effects of the various proposed GUTs, another major one being magnetic monopoles. Both became the focus of major experimental physics efforts starting in the early 1980s. Proton decay was, for a time, an extremely exciting area of experimental physics research. To date, all attempts to observe these events have failed. Recent experiments at the Super-Kamiokande water Cherenkov radiation detector in Japan gave lower limits for proton half-life, at 90% confidence level, of 6.6×10^{33} years via antimuon decay and 8.2×10^{33} years via positron decay.[3] Newer, preliminary results estimate a half-life of no less than 1.29×10^{34} years via positron decay.[4]

A 2014 result with 260kT·yr of data, searching for decay to K-mesons set a lower limit of 5.9×10^{33} yr,[5] close to a supersymmetry (SUSY) prediction of near 10^{34} yr.[6]

7.3 Theoretical motivation

Despite the lack of observational evidence for proton decay, some grand unification theories, such as the Georgi–Glashow model, require it. According to some such theories, the proton has a half-life of about 10^{36} years, and decays into a positron and a neutral pion that itself immediately decays into 2 gamma ray photons:

Since a positron is an antilepton this decay preserves B-L number, which is conserved in most GUTs.

Additional decay modes are available (e.g.: p+ → µ+ + π0),[3] both directly and when catalyzed via interaction with GUT-predicted magnetic monopoles.[7] Though this process has not been observed experimentally, it is within the realm of experimental testability for future planned very large-scale detectors on the megaton scale. Such detectors include the Hyper-Kamiokande.

Early grand unification theories, such as the Georgi–Glashow model, which were the first consistent theories to suggest proton decay postulated that the proton's half-life would be at least 10^{31} years. As further experiments and calculations were performed in the 1990s, it became clear that the proton half-life could not lie below 10^{32} years. Many books from that period refer to this figure for the possible decay time for baryonic matter.

Although the phenomenon is referred to as "proton decay", the effect would also be seen in neutrons bound inside atomic nuclei. Free neutrons—those not inside an atomic nucleus—are already known to decay into protons (and an electron and an antineutrino) in a process called beta decay. Free neutrons have a half-life of about 10 minutes (613.9±0.8 s)[8] due to the weak interaction. Neutrons bound inside a nucleus have an immensely longer half-life—apparently as great as that of the proton.

7.4 Decay operators

7.4.1 Dimension-6 proton decay operators

The dimension-6 proton decay operators are $\frac{qqql}{\Lambda^2}$, $\frac{d^c u^c u^c e^c}{\Lambda^2}$, $\frac{\overline{e^c u^c} qq}{\Lambda^2}$ and $\frac{\overline{d^c u^c} ql}{\Lambda^2}$ where Λ is the cutoff scale for the Standard Model. All of these operators violate both baryon number (B) and lepton number (L) conservation but not the combination $B - L$.

In GUT models, the exchange of an X or Y boson with the mass ΛGUT can lead to the last two operators suppressed by $\frac{1}{\Lambda_{GUT}^2}$. The exchange of a triplet Higgs with mass M can lead to all of the operators suppressed by $1/M^2$. See doublet–triplet splitting problem.

- Proton Decay. These graphics refer to the X bosons and Higgs bosons.

- Dimension 6 proton decay mediated by the X boson (3,2)
  $-\frac{5}{6}$ in SU(5) GUT

- Dimension 6 proton decay mediated by the X boson (3,2)
  $\frac{1}{6}$ in flipped SU(5) GUT

- Dimension 6 proton decay mediated by the triplet Higgs T (3,1)
  $-\frac{1}{3}$ and the anti-triplet Higgs T (3,1)
  $\frac{1}{3}$ in SU(5) GUT

7.4.2 Dimension-5 proton decay operators

In supersymmetric extensions (such as the MSSM), we can also have dimension-5 operators involving two fermions and two sfermions caused by the exchange of a tripletino of mass M. The sfermions will then exchange a gaugino or Higgsino or gravitino leaving two fermions. The overall Feynman diagram has a loop (and other complications due to strong interaction physics). This decay rate is suppressed by $\frac{1}{MM_{SUSY}}$ where MSUSY is the mass scale of the superpartners.

7.4.3   Dimension-4 proton decay operators

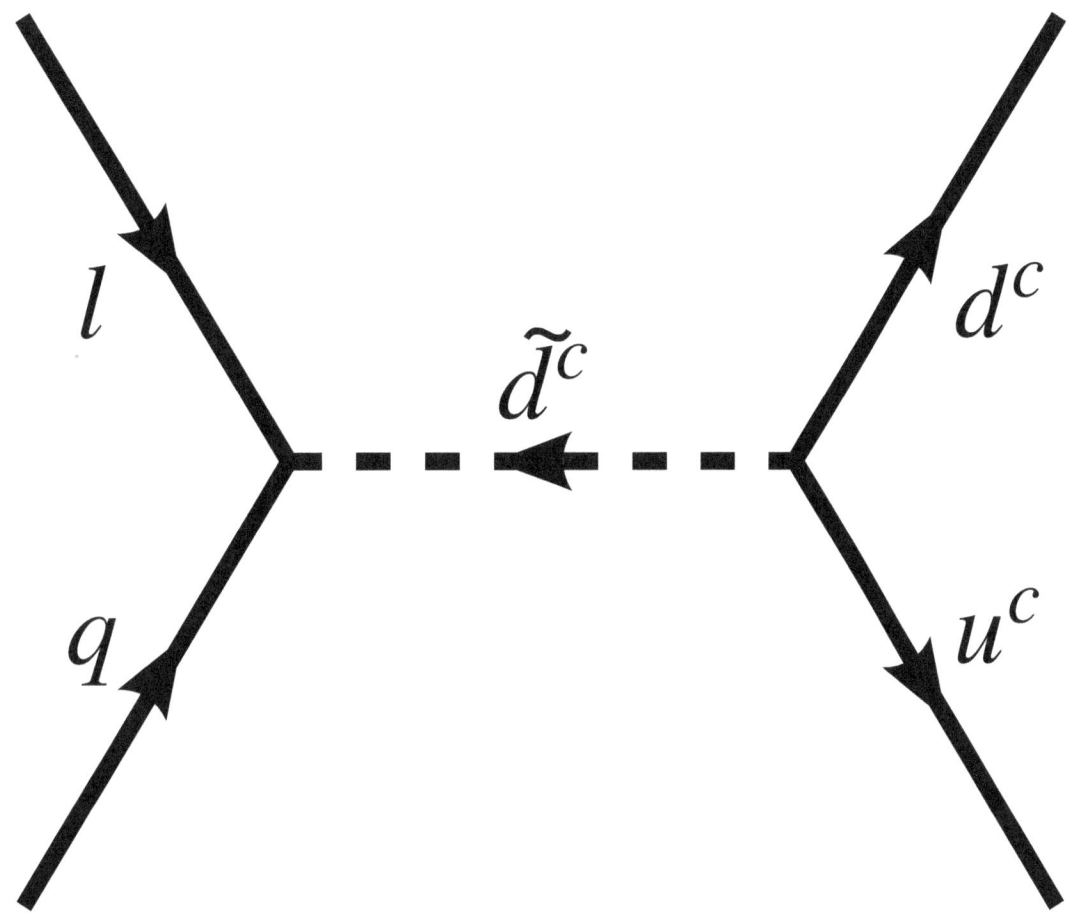

In the absence of matter parity, supersymmetric extensions of the Standard Model can give rise to the last operator suppressed by the inverse square of sdown quark mass. This is due to the dimension-4 operators qld^c and $u^c d^c d^c$.

The proton decay rate is only suppressed by $\frac{1}{M_{SUSY}^2}$ which is far too fast unless the couplings are very small.

7.5   See also

- Virtual black hole

- Weak hypercharge

- B − L

- X and Y bosons

7.6 References

[1] Radioactive decays by Protons. Myth or reality?, Ishfaq Ahmad, The Nucleus, 1969. pp 69-70

[2] V.M. Abazov et al. (2010). "Evidence for an anomalous like-sign dimuon charge asymmetry". arXiv:1005.2757.

[3] H. Nishino; Super-K Collaboration (2012). "Search for Proton Decay via p+ → e+π0 and p+ → μ+π0 in a Large Water Cherenkov Detector".*Physical Review Letters***102**(14):141801.Bibcode:2009PhRvL.102n1801N.doi:10.1103/PhysRevLett.

[4] http://www-sk.icrr.u-tokyo.ac.jp/whatsnew/new-20091125-e.html

[5] K. Abe et al. (Super-Kamiokande Collaboration) (14 October 2014). "Search for proton decay via p→vK+ using 260 kiloton·year data of Super-Kamiokande". *Phys. Rev. D* **90**. Bibcode:2014PhRvD..90g2005A. doi:10.1103/PhysRevD.90.072005.

[6] Schirber, Michael. "Synopsis: Proton Longevity Pushes New Bounds". *Physics*. American Physical Society. Retrieved 20 October 2014.

[7] B. V. Sreekantan (1984). "Searches for Proton Decay and Superheavy Magnetic Monopoles" (PDF). *Journal of Astrophysics and Astronomy* **5** (3): 251–271. Bibcode:1984JApA....5..251S. doi:10.1007/BF02714542.

[8] W.-M. Yao et al. (2006). "Review of Particle Physics – *N* Baryons" (PDF). *Journal of Physics G* **33**: 1. arXiv:astro-ph/0601168. Bibcode:2006JPhG...33....1Y. doi:10.1088/0954-3899/33/1/001.

7.7 Further reading

- C. Amsler; Particle Data Group (2008). "Review of Particle Physics – *N* Baryons" (PDF). *Physics Letters B* **667**: 1. Bibcode:2008PhLB..667....1P. doi:10.1016/j.physletb.2008.07.018.

- K. Hagiwara; Particle Data Group (2002). "Review of Particle Physics – *N* Baryons" (PDF). *Physical Review D* **66**: 010001. Bibcode:2002PhRvD..66a0001H. doi:10.1103/PhysRevD.66.010001.

- F. Adams; G. Laughlin. *The Five Ages of the Universe : Inside the Physics of Eternity*. ISBN 978-0-684-86576-8.

- L.M. Krauss. *Atom : An Odyssey from the Big Bang to Life on Earth*. ISBN 0-316-49946-3.

- D.-D.Wu;T.-Z.Li. "Proton decay,annihilation or fusion?".*Zeitschrift für Physik C***27**(2):321–323.Bibcode:11W. doi:10.1007/BF01556623.

- P. Nath; P. Fileviez Perez (2007). "Proton stability in grand unified theories, in strings and in branes". *Physics Reports***441**(5-6):191–317.arXiv:hep-ph/0601023.Bibcode:2007PhR...441..191N.doi:10.1016/j.physrep.2007.020.

7.8 External links

- Proton decay at Super-Kamiokande

- Pictorial history of the IMB experiment

- L. Maiani (2006). "The problem of proton decay" (PDF). *3rd International Workshop on NO-VE.*

Chapter 8

Radioactive decay

For particle decay in a more general context, see Particle decay. For more information on hazards of various kinds of radiation from decay, see Ionizing radiation.

"Radioactive" redirects here. For other uses, see Radioactive (disambiguation).

"Radioactivity" redirects here. For other uses, see Radioactivity (disambiguation).

Radioactive decay, also known as **nuclear decay** or **radioactivity**, is the process by which a nucleus of an unstable atom

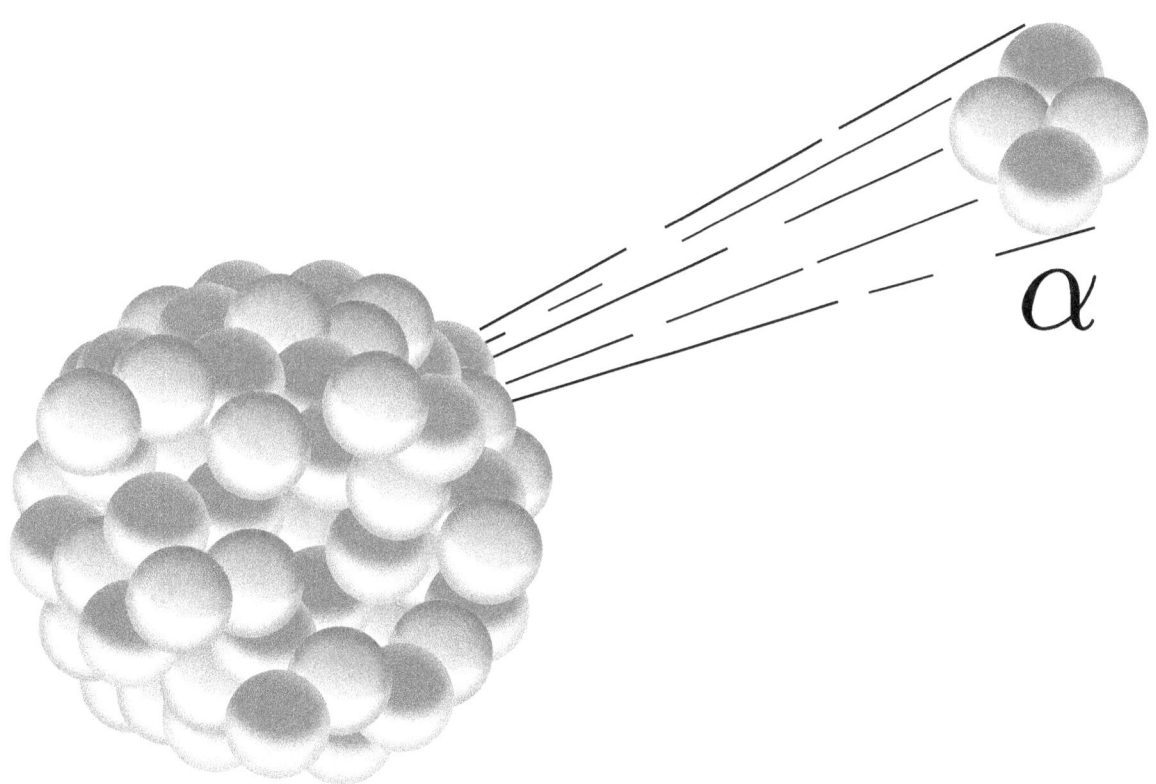

Alpha decay is one type of radioactive decay, in which an atomic nucleus emits an alpha particle, and thereby transforms (or "decays") into an atom with a mass number decreased by 4 and atomic number decreased by 2.

loses energy by emitting radiation. A material that spontaneously emits such radiation — which includes alpha particles, beta particles, gamma rays and conversion electrons — is considered **radioactive**.

Radioactive decay is a stochastic (i.e. random) process at the level of single atoms, in that, according to quantum theory, it is impossible to predict when a particular atom will decay.[1][2][3][4] The chance that a given atom will decay never changes, that is, it does not matter how long the atom has existed. For a large collection of atoms however, the decay rate for that collection can be calculated from their measured decay constants or half-lives. This is the basis of radiometric dating. The half-lives of radioactive atoms have no known limits for shortness or length of duration, and range over 55 orders of magnitude in time.

There are many types of radioactive decay (see table below). A decay, or loss of energy, results when an atom with one type of nucleus, called the *parent radionuclide* (or *parent radioisotope*[note 1]), transforms into an atom with a nucleus in a different state, or with a nucleus containing a different number of protons and neutrons. The product is called the *daughter nuclide*. In some decays, the parent and the daughter nuclides are different chemical elements, and thus the decay process results in the creation of an atom of a different element. This is known as a nuclear transmutation.

The first decay processes to be discovered were alpha decay, beta decay, and gamma decay. Alpha decay occurs when the nucleus ejects an alpha particle (helium nucleus). This is the most common process of emitting nucleons, but in rarer types of decays, nuclei can eject protons, or in the case of cluster decay specific nuclei of other elements. Beta decay occurs when the nucleus emits an electron or positron and a neutrino, in a process that changes a proton to a neutron or the other way about. The nucleus may capture an orbiting electron, causing a proton to convert into a neutron in a process called electron capture. All of these processes result in a nuclear transmutation.

By contrast, there are radioactive decay processes that do not result in a nuclear transmutation. The energy of an excited nucleus may be emitted as a gamma ray in a process called gamma decay, or be used to eject an orbital electron by its interaction with the excited nucleus, in a process called internal conversion. Highly excited neutron-rich nuclei, formed as the product of other types of decay, occasionally lose energy by way of neutron emission, resulting in a change of an element from one isotope to another. Another type of radioactive decay results in products that are not defined, but appear in a range of "pieces" of the original nucleus. This decay, called spontaneous fission, happens when a large unstable nucleus spontaneously splits into two (and occasionally three) smaller daughter nuclei, and generally leads to the emission of gamma rays, neutrons, or other particles from those products.

For a summary table showing the number of stable and radioactive nuclides in each category, see radionuclide. There exist twenty-nine chemical elements on Earth that are radioactive. They are those that contain thirty-four radionuclides that date before the time of formation of the solar system. Well-known examples are uranium and thorium, but also included are naturally occurring long-lived radioisotopes such as potassium-40. Another fifty or so shorter-lived radionuclides, such as radium and radon, found on Earth, are the products of decay chains that began with the primordial nuclides, and ongoing cosmogenic processes, such as the production of carbon-14 from nitrogen-14 by cosmic rays. Radionuclides may also be produced artificially in particle accelerators or nuclear reactors, resulting in 650 of these with half-lives of over an hour, and several thousand more with even shorter half-lives. See this list of nuclides for a list by half life.

8.1 History of discovery

Radioactivity was discovered in 1896 by the French scientist Henri Becquerel, while working with phosphorescent materials.[5] These materials glow in the dark after exposure to light, and he suspected that the glow produced in cathode ray tubes by X-rays might be associated with phosphorescence. He wrapped a photographic plate in black paper and placed various phosphorescent salts on it. All results were negative until he used uranium salts. The uranium salts caused a blackening of the plate in spite of the plate being wrapped in black paper. These radiations were given the name "Becquerel Rays".

It soon became clear that the blackening of the plate had nothing to do with phosphorescence, as the blackening was also produced by non-phosphorescent salts of uranium and metallic uranium. It became clear from these experiments that there was a form of invisible radiation that could pass through paper and was causing the plate to react as if exposed to light.

At first, it seemed as though the new radiation was similar to the then recently discovered X-rays. Further research by Becquerel, Ernest Rutherford, Paul Villard, Pierre Curie, Marie Curie, and others showed that this form of radioactivity was significantly more complicated. Rutherford was the first to realize that all such elements decay in accordance with the same mathematical exponential formula. Rutherford and his student Frederick Soddy were the first to realize that many decay processes resulted in the transmutation of one element to another. Subsequently, the radioactive displacement law

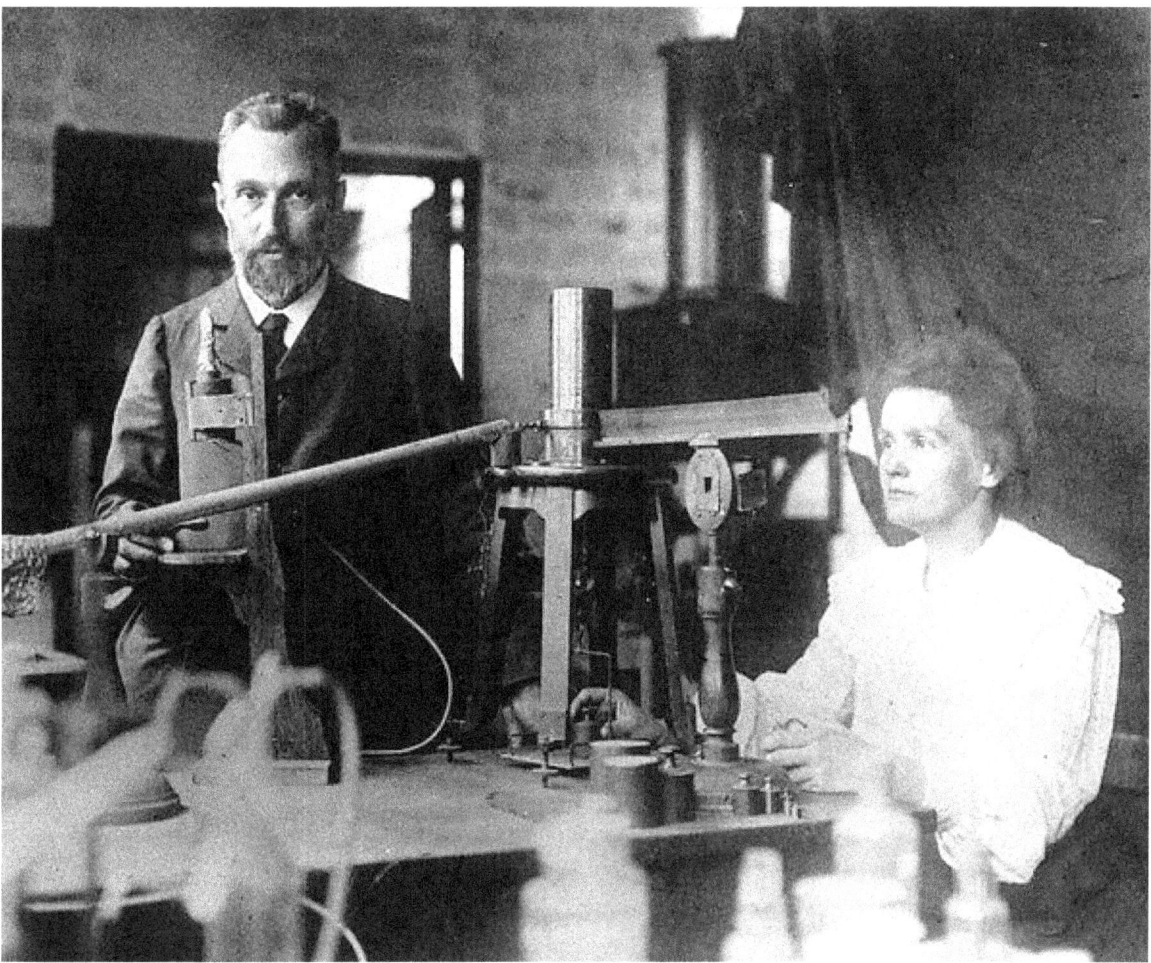

Pierre and Marie Curie in their Paris laboratory, before 1907

of Fajans and Soddy was formulated to describe the products of alpha and beta decay.[6][7]

The early researchers also discovered that many other chemical elements, besides uranium, have radioactive isotopes. A systematic search for the total radioactivity in uranium ores also guided Pierre and Marie Curie to isolate two new elements: polonium and radium. Except for the radioactivity of radium, the chemical similarity of radium to barium made these two elements difficult to distinguish.

8.2   Early health dangers

The dangers of ionizing radiation due to radioactivity and X-rays were not immediately recognized.

8.2.1   X-rays

The discovery of x-rays by Wilhelm Röntgen in 1895 led to widespread experimentation by scientists, physicians, and inventors. Many people began recounting stories of burns, hair loss and worse in technical journals as early as 1896. In February of that year, Professor Daniel and Dr. Dudley of Vanderbilt University performed an experiment involving X-raying Dudley's head that resulted in his hair loss. A report by Dr. H.D. Hawks, of his suffering severe hand and chest burns in an X-ray demonstration, was the first of many other reports in *Electrical Review*.[8]

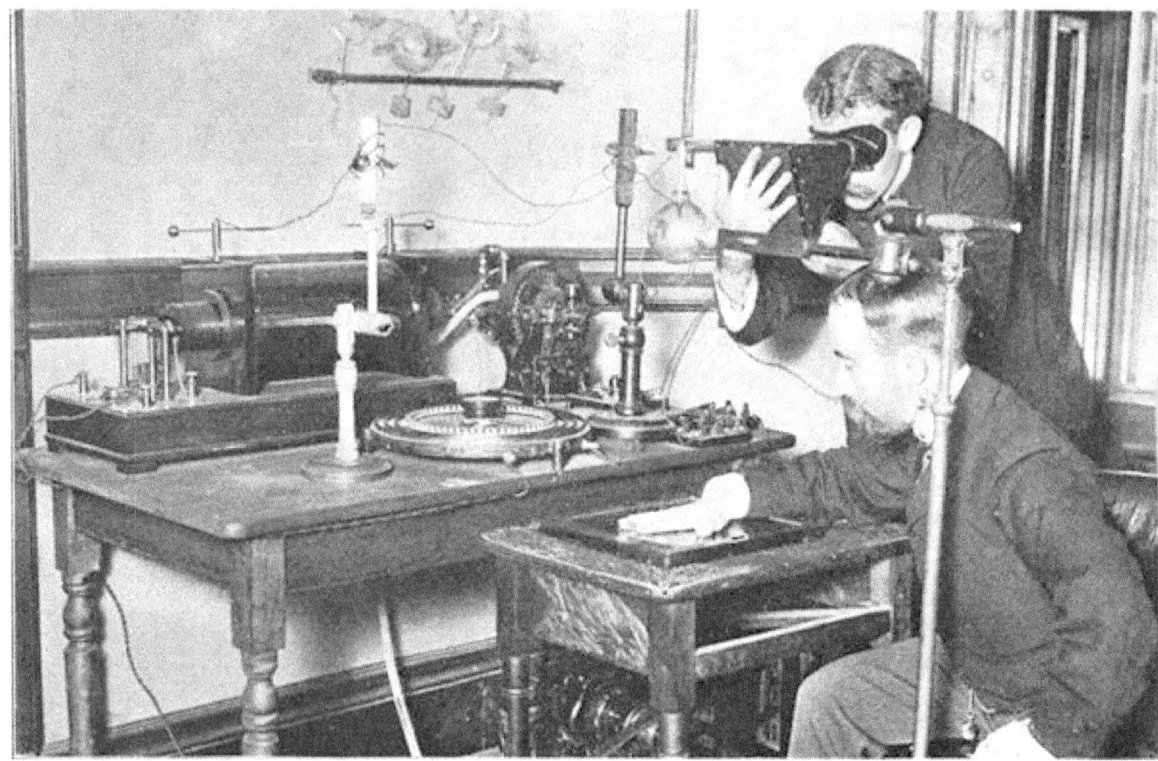

Taking an X-ray image with early Crookes tube apparatus in 1896. The Crookes tube is visible in the centre. The standing man is viewing his hand with a fluoroscope screen; this was a common way of setting up the tube. No precautions against radiation exposure are being taken; its hazards were not known at the time.

Other experimenters including Elihu Thomson, and Nikola Tesla also reported burns. Thomson deliberately exposed a finger to an X-ray tube over a period of time and suffered pain, swelling, and blistering.[9] Other effects, including ultraviolet rays and ozone were sometimes blamed for the damage,[10] and many physicians still claimed that there were no effects from X-ray exposure at all.[9]

Despite this, there were some early systematic hazard investigations, and as early as 1902 William Herbert Rollins wrote almost despairingly that his warnings about the dangers involved in careless use of X-rays was not being heeded, either by industry or by his colleagues. By this time Rollins had proved that X-rays could kill experimental animals, could cause a pregnant guinea pig to abort, and that they could kill a fetus.[11] He also stressed that "animals vary in susceptibility to the external action of X-light" and warned that these differences be considered when patients were treated by means of X-rays.

8.2.2 Radioactive substances

However, the biological effects of radiation due to radioactive substances were less easy to gauge. This gave the opportunity for many physicians and corporations to market radioactive substances as patent medicines. Examples were radium enema treatments, and radium-containing waters to be drunk as tonics. Marie Curie protested against this sort of treatment, warning that the effects of radiation on the human body were not well understood. Curie later died from aplastic anaemia, likely caused by exposure to ionizing radiation. By the 1930s, after a number of cases of bone necrosis and death of radium treatment enthusiasts, radium-containing medicinal products had been largely removed from the market (radioactive quackery).

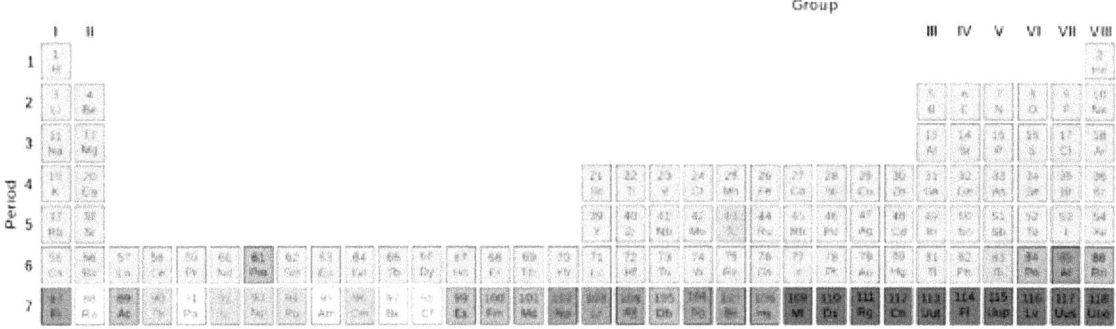

Radioactivity is characteristic of elements with large atomic number. Elements with at least one stable isotope are shown in light blue. Green shows elements whose most stable isotope has a half-life measured in millions of years. Yellow and orange are progressively more unstable, with half-lives in thousands or hundreds of years, down toward one day. Red and purple show highly and extremely radioactive elements where the most stable isotopes exhibit half-lives measured on the order of one day and much less.

8.2.3 Radiation protection

Main article: Radiation protection
See also: Sievert and Ionizing radiation

Only a year after Röntgen's discovery of X rays, the American engineer Wolfram Fuchs (1896) gave what is probably the first protection advice, but it was not until 1925 that the first International Congress of Radiology (ICR) was held and considered establishing international protection standards. The effects of radiation on genes, including the effect of cancer risk, were recognized much later. In 1927, Hermann Joseph Muller published research showing genetic effects and, in 1946, was awarded the Nobel prize for his findings.

The second ICR was held in Stockholm in 1928 and proposed the adoption of the rontgen unit, and the 'International X-ray and Radium Protection Committee' (IXRPC) was formed. Rolf Sievert was named Chairman, but a driving force was George Kaye of the British National Physical Laboratory. The committee met in 1931, 1934 and 1937.

After World War II the increased range and quantity of radioactive substances being handled as a result of military and civil nuclear programmes led to large groups of occupational workers and the public being potentially exposed to harmful levels of ionising radiation. This was considered at the first post-war ICR convened in London in 1950, when the present International Commission on Radiological Protection (ICRP) was born.[12] Since then the ICRP has developed the present international system of radiation protection, covering all aspects of radiation hazard.

8.3 Units of radioactivity

The International System of Units (SI) unit of radioactive activity is the becquerel (Bq), named in honour of the scientist Henri Becquerel. One Bq is defined as one transformation (or decay or disintegration) per second.

An older unit of radioactivity is the curie, Ci, which was originally defined as "the quantity or mass of radium emanation in equilibrium with one gram of radium (element)".[13] Today, the curie is defined as 3.7×10^{10} disintegrations per second, so that 1 curie (Ci) = 3.7×10^{10} Bq. For radiological protection purposes, although the United States Nuclear Regulatory Commission permits the use of the unit curie alongside SI units,[14] the European Union European units of measurement directives required that its use for "public health ... purposes" be phased out by 31 December 1985.[15]

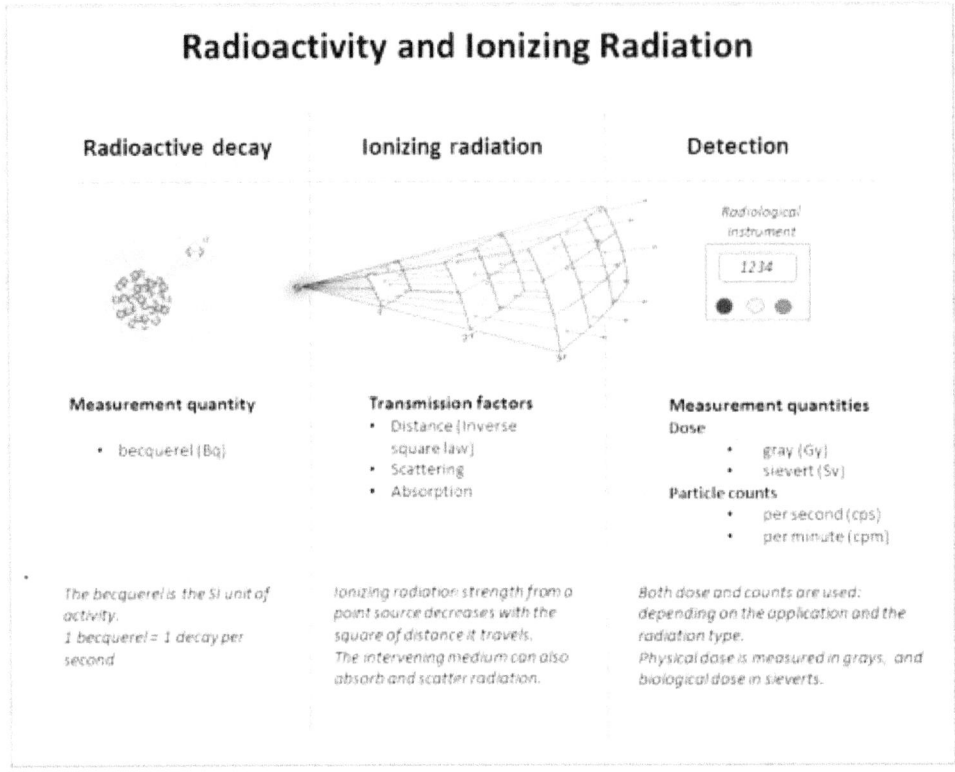

Graphic showing relationships between radioactivity and detected ionizing radiation

8.4 Types of decay

Early researchers found that an electric or magnetic field could split radioactive emissions into three types of beams. The rays were given the names alpha, beta, and gamma, in order of their ability to penetrate matter. While alpha decay was seen only in heavier elements of atomic number 52 (tellurium) and greater, the other two types of decay were produced by all of the elements. Lead, atomic number 82, is the heaviest element to have any isotopes stable (to the limit of measurement) to radioactive decay. Radioactive decay is seen in all isotopes of all elements of atomic number 83 (bismuth) or greater. Bismuth, however, is only very slightly radioactive.

In analysing the nature of the decay products, it was obvious from the direction of the electromagnetic forces applied to the radiations by external magnetic and electric fields that alpha particles carried a positive charge, beta particles carried a negative charge, and gamma rays were neutral. From the magnitude of deflection, it was clear that alpha particles were much more massive than beta particles. Passing alpha particles through a very thin glass window and trapping them in a discharge tube allowed researchers to study the emission spectrum of the captured particles, and ultimately proved that alpha particles are helium nuclei. Other experiments showed beta radiation, resulting from decay and cathode rays, were high-speed electrons. Likewise, gamma radiation and X-rays were found to be high-energy electromagnetic radiation.

The relationship between the types of decays also began to be examined: For example, gamma decay was almost always found to be associated with other types of decay, and occurred at about the same time, or afterwards. Gamma decay as a separate phenomenon, with its own half-life (now termed isomeric transition), was found in natural radioactivity to be a result of the gamma decay of excited metastable nuclear isomers, which were in turn created from other types of decay.

Although alpha, beta, and gamma radiations were most commonly found, other types of emission were eventually discovered. Shortly after the discovery of the positron in cosmic ray products, it was realized that the same process that

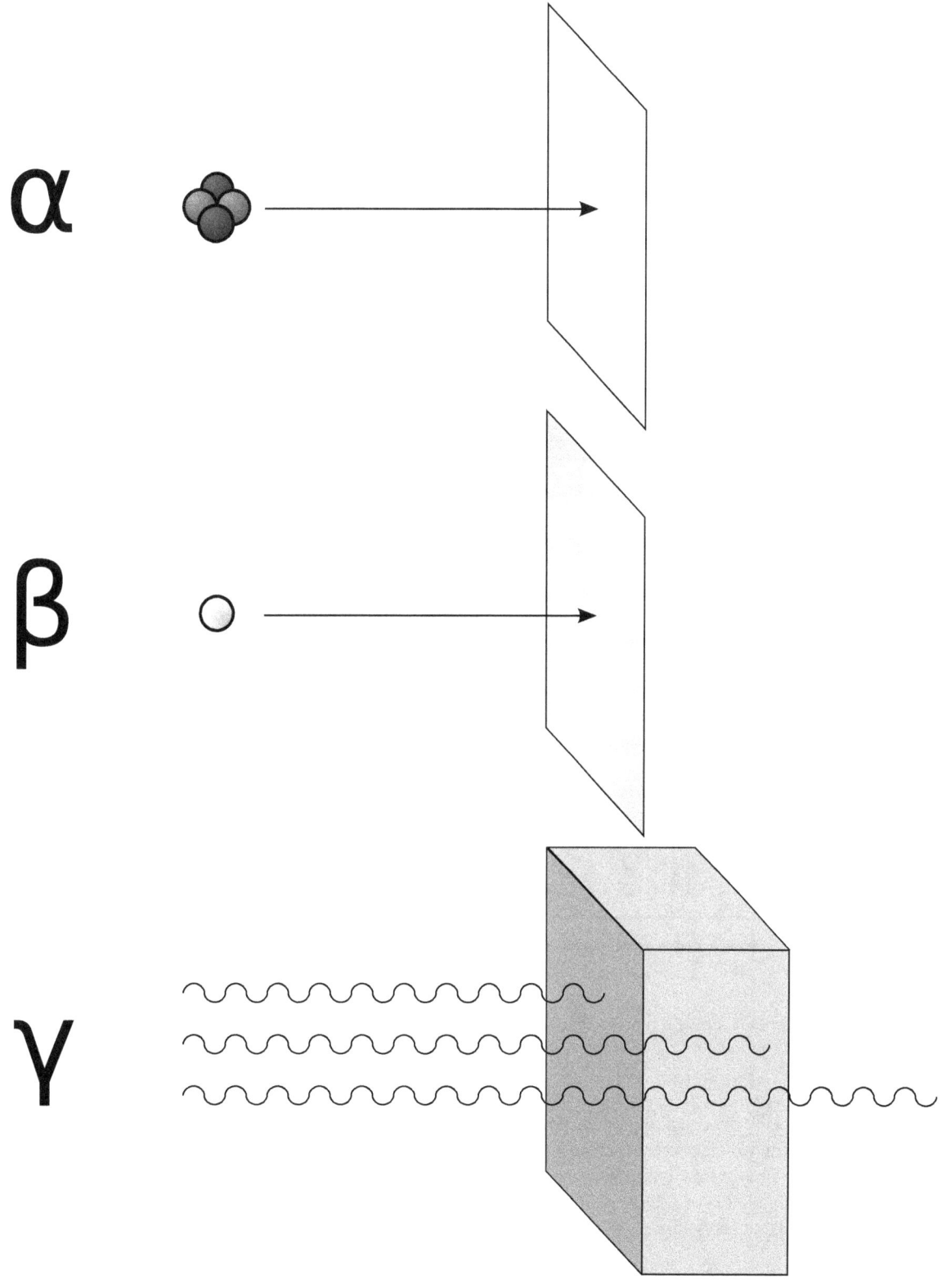

Alpha particles may be completely stopped by a sheet of paper, beta particles by aluminium shielding. Gamma rays can only be reduced by much more substantial mass, such as a very thick layer of lead.

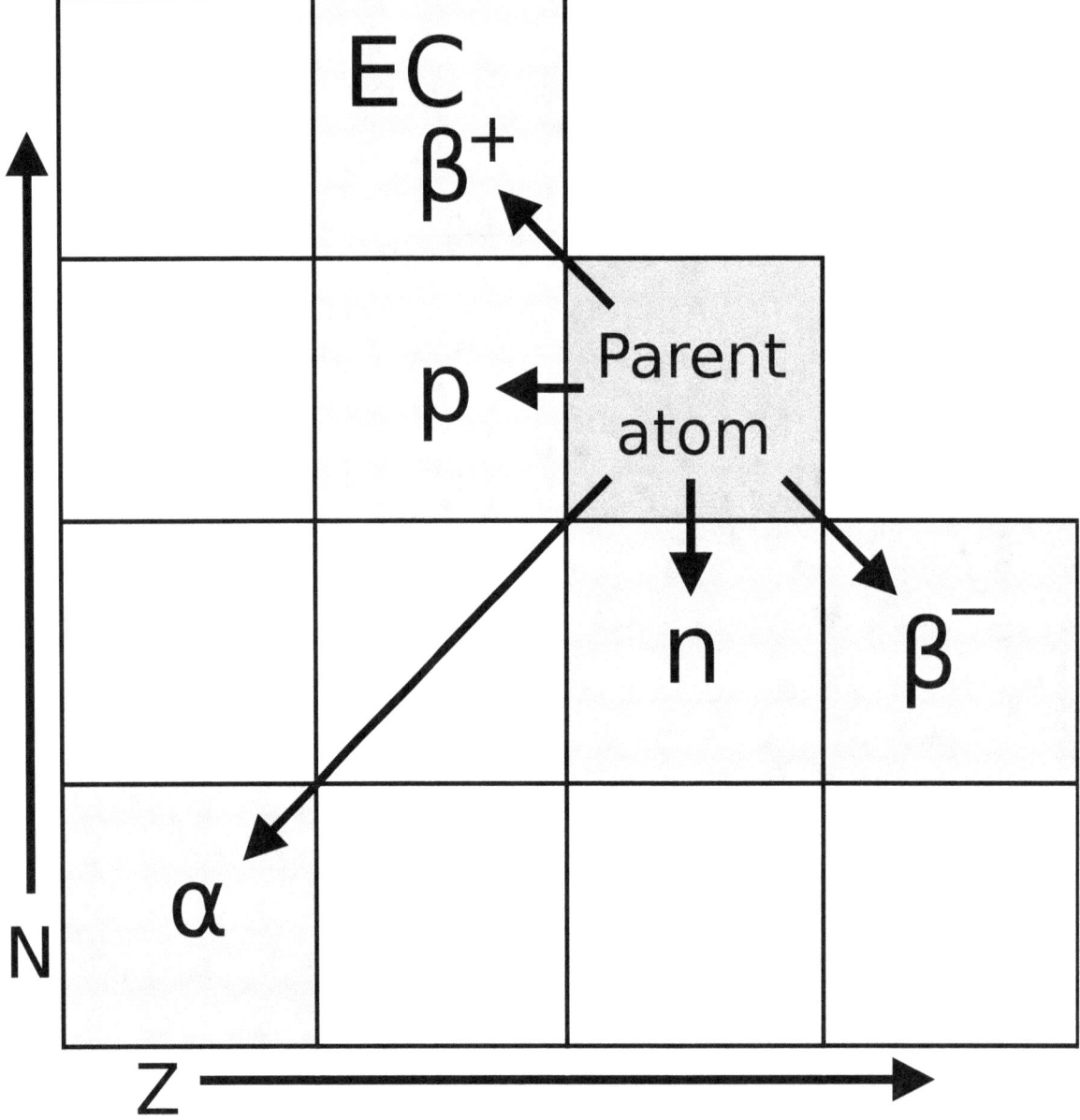

Transition diagram for decay modes of a radionuclide, with neutron number N and atomic number Z (shown are α, $\beta^{\pm}$, p^+, and n^0 emissions, EC denotes electron capture).

operates in classical beta decay can also produce positrons (positron emission). In an analogous process, instead of emitting positrons and neutrinos, some proton-rich nuclides were found to capture their own atomic electrons, a process called electron capture, and subsequently emit only a neutrino and usually also a gamma ray. Each of these types of decay involves the capture or emission of nuclear electrons or positrons, and acts to move a nucleus toward the ratio of neutrons to protons that has the least energy for a given total number of nucleons, consequently producing a more stable nucleus.

A theoretical process of positron capture, analogous to electron capture, is possible in antimatter atoms, but has not been observed as antimatter atoms are rarely available.[16] This would require antimatter atoms at least as complex as beryllium-7, which is the lightest known isotope of normal matter to undergo decay by electron capture.

Shortly after the discovery of the neutron in 1932, Enrico Fermi realized that certain rare beta-decay reactions immediately yield neutrons as a decay particle (neutron emission). Isolated proton emission was eventually observed in some elements. It was also found that some heavy elements may undergo spontaneous fission into products that vary in composition. In a phenomenon called cluster decay, specific combinations of neutrons and protons other than alpha particles (helium nuclei) were found to be spontaneously emitted from atoms.

Other types of radioactive decay of different mechanisms were found to emit previously-seen particles. An example is internal conversion, which results in electron and sometimes high-energy photon emission, although it involves neither beta nor gamma decay. A neutrino is not emitted, and neither the electron nor photon originate in the nucleus. Internal conversion decay, like isomeric transition gamma decay and neutron emission, involves the release of energy by an excited nuclide, without the transmutation of one element into another.

Rare events that involve a combination of two beta-decay type events happening simultaneously are known (see below). Any decay process that does not violate the conservation of energy or momentum laws (and perhaps other particle conservation laws) is permitted to happen, although not all have been detected. An interesting example discussed in a final section, is bound state beta decay of rhenium-187. In this process, an inverse of electron capture, beta electron-decay of the parent nuclide is not accompanied by beta electron emission, because the beta particle has been captured into the K-shell of the emitting atom. An antineutrino, however, is emitted.

Radionuclides can undergo a number of different reactions. These are summarized in the following table. A nucleus with mass number A and atomic number Z is represented as (A, Z). The column "Daughter nucleus" indicates the difference between the new nucleus and the original nucleus. Thus, $(A - 1, Z)$ means that the mass number is one less than before, but the atomic number is the same as before.

If energy circumstances are favorable, a given radionuclide may undergo many competing types of decay, with some atoms decaying by one route, and others decaying by another. An example is copper-64, which has 29 protons, and 35 neutrons, which decays with a half-life of about 12.7 hours. This isotope has one unpaired proton and one unpaired neutron, so either the proton or the neutron can decay to the opposite particle. This particular nuclide (though not all nuclides in this situation) is almost equally likely to decay through proton decay, producing a positron emission (18%), or through electron capture (43%), as it does through neutron decay by electron emission (39%). The excited energy states resulting from these decays which fail to end in a ground energy state, also produce later internal conversion and gamma decay in almost 0.5% of the time.

Radioactive decay results in a reduction of summed rest mass, once the released energy (the *disintegration energy*) has escaped in some way. Although decay energy is sometimes defined as associated with the difference between the mass of the parent nuclide products and the mass of the decay products, this is true only of rest mass measurements, where some energy has been removed from the product system. This is true because the decay energy must always carry mass with it, wherever it appears (see mass in special relativity) according to the formula $E = mc^2$. The decay energy is initially released as the energy of emitted photons plus the kinetic energy of massive emitted particles (that is, particles that have rest mass). If these particles come to thermal equilibrium with their surroundings and photons are absorbed, then the decay energy is transformed to thermal energy, which retains its mass.

Decay energy therefore remains associated with a certain measure of mass of the decay system, called invariant mass, which does not change during the decay, even though the energy of decay is distributed among decay particles. The energy of photons, the kinetic energy of emitted particles, and, later, the thermal energy of the surrounding matter, all contribute to the invariant mass of the system. Thus, while the sum of the rest masses of the particles is not conserved in radioactive decay, the *system* mass and system invariant mass (and also the system total energy) is conserved throughout any decay process. This is a restatement of the equivalent laws of conservation of energy and conservation of mass.

8.5 Radioactive decay rates

The *decay rate*, or *activity*, of a radioactive substance is characterized by:

Constant quantities:

- The *half-life*—$t_1/2$, is the time taken for the activity of a given amount of a radioactive substance to decay to half of its initial value; see List of nuclides.

- The *decay constant*— λ, "lambda" the inverse of the mean lifetime, sometimes referred to as simply *decay rate*.

- The *mean lifetime*— τ, "tau" the average lifetime of a radioactive particle before decay.

Although these are constants, they are associated with the statistical behavior of populations of atoms. In consequence, predictions using these constants are less accurate for minuscule samples of atoms.

In principle a half-life, a third-life, or even a $(1/\sqrt{2})$-life, can be used in exactly the same way as half-life; but the mean life and half-life $t_{1/2}$ have been adopted as standard times associated with exponential decay.

Time-variable quantities:

- *Total activity*— A, is the number of decays per unit time of a radioactive sample.

- *Number of particles*—N, is the total number of particles in the sample.

- *Specific activity*—SA, number of decays per unit time per amount of substance of the sample at time set to zero ($t = 0$). "Amount of substance" can be the mass, volume or moles of the initial sample.

These are related as follows:

$$t_{1/2} = \frac{\ln(2)}{\lambda} = \tau \ln(2)$$

$$A = -\frac{dN}{dt} = \lambda N$$

$$S_A a_0 = -\frac{dN}{dt}\bigg|_{t=0} = \lambda N_0$$

where N_0 is the initial amount of active substance — substance that has the same percentage of unstable particles as when the substance was formed.

8.6 Mathematics of radioactive decay

For the mathematical details of exponential decay in general context, see exponential decay.
For related derivations with some further details, see half-life.
For the analogous mathematics in 1st order chemical reactions, see Consecutive reactions.

8.6.1 Universal law of radioactive decay

Radioactivity is one very frequently given example of exponential decay. The law describes the statistical behaviour of a large number of nuclides, rather than individual atoms. In the following formalism, the number of nuclides or the nuclide population N, is of course a discrete variable (a natural number)—but for any physical sample N is so large that it can be treated as a continuous variable. Differential calculus is needed to set up differential equations for the modelling the behaviour of the nuclear decay.

The mathematics of radioactive decay depend on a key assumption that a nucleus of a radionuclide has no "memory" or way of translating its history into its present behavior. A nucleus does not "age" with the passage of time. Thus, the probability of its breaking down does not increase with time, but stays constant no matter how long the nucleus has existed. This constant probability may vary greatly between different types of nuclei, leading to the many different observed decay rates. However, whatever the probability is, it does not change. This is in marked contrast to complex objects which do show aging, such as automobiles and humans. These systems do have a chance of breakdown per unit of time, that increases from the moment they begin their existence.

One-decay process

Consider the case of a nuclide A that decays into another B by some process $A \to B$ (emission of other particles, like electron neutrinos ν
e and electrons e^- as in beta decay, are irrelevant in what follows). The decay of an unstable nucleus is entirely random and it is impossible to predict when a particular atom will decay.[1] However, it is equally likely to decay at any instant in time. Therefore, given a sample of a particular radioisotope, the number of decay events $-dN$ expected to occur in a small interval of time dt is proportional to the number of atoms present N, that is[17]

$$-\frac{dN}{dt} \propto N.$$

Particular radionuclides decay at different rates, so each has its own decay constant λ. The expected decay $-dN/N$ is proportional to an increment of time, dt:

The negative sign indicates that N decreases as time increases, as the decay events follow one after another. The solution to this first-order differential equation is the function:

$$N(t) = N_0 \, e^{-\lambda t} = N_0 \, e^{-t/\tau},$$

where N_0 is the value of N at time $t = 0$.[17]

We have for all time t:

$$N_A + N_B = N_{\text{total}} = N_{A0},$$

where N_{total} is the constant number of particles throughout the decay process, which is equal to the initial number of A nuclides since this is the initial substance.

If the number of non-decayed A nuclei is:

$$N_A = N_{A0} e^{-\lambda t}$$

then the number of nuclei of B, i.e. the number of decayed A nuclei, is

$$N_B = N_{A0} - N_A = N_{A0} - N_{A0} e^{-\lambda t} = N_{A0} \left(1 - e^{-\lambda t} \right).$$

The number of decays observed over a given interval obeys Poisson statistics. If the average number of decays is $\langle N \rangle$, the probability of a given number of decays N is[17]

$$P(N) = \frac{\langle N \rangle^N \exp(-\langle N \rangle)}{N!}.$$

Chain-decay processes

Chain of two decays

Now consider the case of a chain of two decays: one nuclide A decaying into another B by one process, then B decaying into another C by a second process, i.e. $A \rightarrow B \rightarrow C$. The previous equation cannot be applied to the decay chain, but can be generalized as follows. Since A decays into B, *then* B decays into C, the activity of A adds to the total number of B nuclides in the present sample, *before* those B nuclides decay and reduce the number of nuclides leading to the later sample. In other words, the number of second generation nuclei B increases as a result of the first generation nuclei decay of A, and decreases as a result of its own decay into the third generation nuclei C.[18] The sum of these two terms gives the law for a decay chain for two nuclides:

$$\frac{\mathrm{d}N_B}{\mathrm{d}t} = -\lambda_B N_B + \lambda_A N_A.$$

The rate of change of NB, that is $\mathrm{d}NB/\mathrm{d}t$, is related to the changes in the amounts of A and B, NB can increase as B is produced from A and decrease as B produces C.

Re-writing using the previous results:

The subscripts simply refer to the respective nuclides, i.e. NA is the number of nuclides of type A, NA_0 is the initial number of nuclides of type A, λA is the decay constant for A - and similarly for nuclide B. Solving this equation for NB gives:

$$N_B = \frac{N_{A0}\lambda_A}{\lambda_B - \lambda_A}\left(e^{-\lambda_A t} - e^{-\lambda_B t}\right).$$

In the case where B is a stable nuclide ($\lambda B = 0$), this equation reduces to the previous solution:

$$\lim_{\lambda_B \to 0}\left[\frac{N_{A0}\lambda_A}{\lambda_B - \lambda_A}\left(e^{-\lambda_A t} - e^{-\lambda_B t}\right)\right] = \frac{N_{A0}\lambda_A}{0 - \lambda_A}\left(e^{-\lambda_A t} - 1\right) = N_{A0}\left(1 - e^{-\lambda_A t}\right),$$

as shown above for one decay. The solution can be found by the integration factor method, where the integrating factor is $e^{\lambda_B t}$. This case is perhaps the most useful, since it can derive both the one-decay equation (above) and the equation for multi-decay chains (below) more directly.

Chain of any number of decays

For the general case of any number of consecutive decays in a decay chain, i.e. $A_1 \rightarrow A_2 \cdots \rightarrow Ai \cdots \rightarrow AD$, where D is the number of decays and i is a dummy index ($i = {}_1, {}_2, {}_3, ...D$), each nuclide population can be found in terms of the previous population. In this case $N_2 = 0$, $N_3 = 0$,..., $ND = 0$. Using the above result in a recursive form:

$$\frac{\mathrm{d}N_j}{\mathrm{d}t} = -\lambda_j N_j + \lambda_{j-1} N_{(j-1)0}e^{-\lambda_{j-1}t}.$$

The general solution to the recursive problem is given by ***Bateman's equations***:[19]

Alternative decay modes

In all of the above examples, the initial nuclide decays into only one product.[20] Consider the case of one initial nuclide that can decay into either of two products, that is $A \rightarrow B$ and $A \rightarrow C$ in parallel. For example, in a sample of potassium-40, 89.3% of the nuclei decay to calcium-40 and 10.7% to argon-40. We have for all time t:

$$N = N_A + N_B + N_C$$

which is constant, since the total number of nuclides remains constant. Differentiating with respect to time:

$$\frac{dN_A}{dt} = -\left(\frac{dN_B}{dt} + \frac{dN_C}{dt}\right)$$
$$-\lambda N_A = -N_A\left(\lambda_B + \lambda_C\right)$$

defining the *total decay constant* λ in terms of the sum of *partial decay constants* λB and λC:

$$\lambda = \lambda_B + \lambda_C.$$

Notice that

$$\frac{dN_A}{dt} < 0, \frac{dN_B}{dt} > 0, \frac{dN_C}{dt} > 0.$$

Solving this equation for *NA*:

$$N_A = N_{A0}e^{-\lambda t}.$$

where NA_0 is the initial number of nuclide A. When measuring the production of one nuclide, one can only observe the total decay constant λ. The decay constants λB and λC determine the probability for the decay to result in products *B* or *C* as follows:

$$N_B = \frac{\lambda_B}{\lambda}N_{A0}\left(1 - e^{-\lambda t}\right),$$

$$N_C = \frac{\lambda_C}{\lambda}N_{A0}\left(1 - e^{-\lambda t}\right).$$

because the fraction $\lambda B/\lambda$ of nuclei decay into *B* while the fraction $\lambda C/\lambda$ of nuclei decay into *C*.

8.6.2   Corollaries of the decay laws

The above equations can also be written using quantities related to the number of nuclide particles *N* in a sample;

- The activity: $A = \lambda N$.

- The amount of substance: $n = N/L$.

- The mass: $M = Arn = ArN/L$.

where $L = 6.022 \times 10^{23}$ is Avogadro's constant, *Ar* is the relative atomic mass number, and the amount of the substance is in moles.

8.6.3 Decay timing: definitions and relations

Time constant and mean-life

For the one-decay solution $A \rightarrow B$:

$$N = N_0 \, e^{-\lambda t} = N_0 \, e^{-t/\tau},$$

the equation indicates that the decay constant λ has units of t^{-1}, and can thus also be represented as $1/\tau$, where τ is a characteristic time of the process called the *time constant*.

In a radioactive decay process, this time constant is also the mean lifetime for decaying atoms. Each atom "lives" for a finite amount of time before it decays, and it may be shown that this mean lifetime is the arithmetic mean of all the atoms' lifetimes, and that it is τ, which again is related to the decay constant as follows:

$$\tau = \frac{1}{\lambda}.$$

This form is also true for two-decay processes simultaneously $A \rightarrow B + C$, inserting the equivalent values of decay constants (as given above)

$$\lambda = \lambda_B + \lambda_C$$

into the decay solution leads to:

$$\frac{1}{\tau} = \lambda = \lambda_B + \lambda_C = \frac{1}{\tau_B} + \frac{1}{\tau_C}$$

Half-life

A more commonly used parameter is the half-life. Given a sample of a particular radionuclide, the half-life is the time taken for half the radionuclide's atoms to decay. For the case of one-decay nuclear reactions:

$$N = N_0 \, e^{-\lambda t} = N_0 \, e^{-t/\tau},$$

the half-life is related to the decay constant as follows: set $N = N_0/2$ and $t = T_1/2$ to obtain

$$t_{1/2} = \frac{\ln 2}{\lambda} = \tau \ln 2.$$

This relationship between the half-life and the decay constant shows that highly radioactive substances are quickly spent, while those that radiate weakly endure longer. Half-lives of known radionuclides vary widely, from more than 10^{19} years, such as for the very nearly stable nuclide ^{209}Bi, to 10^{-23} seconds for highly unstable ones.

The factor of $\ln(2)$ in the above relations results from the fact that concept of "half-life" is merely a way of selecting a different base other than the natural base e for the lifetime expression. The time constant τ is the $e-1$ -life, the time until only $1/e$ remains, about 36.8%, rather than the 50% in the half-life of a radionuclide. Thus, τ is longer than $t_1/2$. The following equation can be shown to be valid:

$$N(t) = N_0 \, e^{-t/\tau} = N_0 \, 2^{-t/t_{1/2}}.$$

Since radioactive decay is exponential with a constant probability, each process could as easily be described with a different constant time period that (for example) gave its "(1/3)-life" (how long until only 1/3 is left) or "(1/10)-life" (a time period until only 10% is left), and so on. Thus, the choice of τ and $t1/2$ for marker-times, are only for convenience, and from convention. They reflect a fundamental principle only in so much as they show that the *same proportion* of a given radioactive substance will decay, during any time-period that one chooses.

Mathematically, the nth life for the above situation would be found in the same way as above—by setting $N = N_0/n$, {{{1}}} and substituting into the decay solution to obtain

$$t_{1/n} = \frac{\ln n}{\lambda} = \tau \ln n.$$

8.6.4   Example

A sample of ^{14}C has a half-life of 5,730 years and a decay rate of 14 disintegration per minute (dpm) per gram of natural carbon.

If an artifact is found to have radioactivity of 4 dpm per gram of its present C, we can find the approximate age of the object using the above equation:

$$N = N_0 \, e^{-t/\tau},$$

where: $\frac{N}{N_0} = 4/14 \approx 0.286$,

$$\tau = \frac{T_{1/2}}{\ln 2} \approx 8267$$

$$t = -\tau \, \ln \frac{N}{N_0} \approx 10360$$

8.7   Changing decay rates

The radioactive decay modes of electron capture and internal conversion are known to be slightly sensitive to chemical and environmental effects that change the electronic structure of the atom, which in turn affects the presence of **1s** and **2s** electrons that participate in the decay process. A small number of mostly light nuclides are affected. For example, chemical bonds can affect the rate of electron capture to a small degree (in general, less than 1%) depending on the proximity of electrons to the nucleus. In ^{7}Be, a difference of 0.9% has been observed between half-lives in metallic and insulating environments.[21] This relatively large effect is because beryllium is a small atom whose valence electrons are in **2s** atomic orbitals, which are subject to electron capture in ^{7}Be because (like all **s** atomic orbitals in all atoms) they naturally penetrate into the nucleus.

In 1992, Jung et al. of the Darmstadt Heavy-Ion Research group observed an accelerated β decay of ^{163}Dy^{66+}. Although neutral ^{163}Dy is a stable isotope, the fully ionized ^{163}Dy^{66+} undergoes β decay into the K and L shells with a half-life of 47 days.[22]

Rhenium-187 is another spectacular example. ^{187}Re normally beta decays to ^{187}Os with a half-life of 41.6×10^9 years,[23] but studies using fully ionised ^{187}Re atoms (bare nuclei) have found that this can decrease to only 33 years. This is attributed to "bound-state β⁻ decay" of the fully ionised atom – the electron is emitted into the "K-shell" (**1s** atomic orbital), which cannot occur for neutral atoms in which all low-lying bound states are occupied.[24]

A number of experiments have found that decay rates of other modes of artificial and naturally occurring radioisotopes are, to a high degree of precision, unaffected by external conditions such as temperature, pressure, the chemical environment, and electric, magnetic, or gravitational fields.[25] Comparison of laboratory experiments over the last century, studies of the Oklo natural nuclear reactor (which exemplified the effects of thermal neutrons on nuclear decay), and astrophysical observations of the luminosity decays of distant supernovae (which occurred far away so the light has taken a great deal of time to reach us), for example, strongly indicate that unperturbed decay rates have been constant (at least to within the limitations of small experimental errors) as a function of time as well.

Recent results suggest the possibility that decay rates might have a weak dependence on environmental factors. It has been suggested that measurements of decay rates of silicon-32, manganese-54, and radium-226 exhibit small seasonal variations (of the order of 0.1%),[26][27][28] while the decay of Radon-222 exhibit large 4% peak-to-peak seasonal variations,[29] proposed to be related to either solar flare activity or distance from the Sun. However, such measurements are highly susceptible to systematic errors, and a subsequent paper[30] has found no evidence for such correlations in seven other isotopes (^{22}Na, ^{44}Ti, ^{108}Ag, ^{121}Sn, ^{133}Ba, ^{241}Am, ^{238}Pu), and sets upper limits on the size of any such effects.

8.8 Theoretical basis of decay phenomena

The neutrons and protons that constitute nuclei, as well as other particles that approach close enough to them, are governed by several interactions. The strong nuclear force, not observed at the familiar macroscopic scale, is the most powerful force over subatomic distances. The electrostatic force is almost always significant, and, in the case of beta decay, the weak nuclear force is also involved.

The interplay of these forces produces a number of different phenomena in which energy may be released by rearrangement of particles in the nucleus, or else the change of one type of particle into others. These rearrangements and transformations may be hindered energetically, so that they do not occur immediately. In certain cases, random quantum vacuum fluctuations are theorized to promote relaxation to a lower energy state (the "decay") in a phenomenon known as quantum tunneling. Radioactive decay half-life of nuclides has been measured over timescales of 55 orders of magnitude, from 2.3×10^{-23} seconds (for hydrogen-7) to 6.9×10^{31} seconds (for tellurium-128).[31] The limits of these timescales are set by the sensitivity of instrumentation only, and there are no known natural limits to how brief or long a decay half life for radioactive decay of a radionuclide may be.

The decay process, like all hindered energy transformations, may be analogized by a snowfield on a mountain. While friction between the ice crystals may be supporting the snow's weight, the system is inherently unstable with regard to a state of lower potential energy. A disturbance would thus facilitate the path to a state of greater entropy: The system will move towards the ground state, producing heat, and the total energy will be distributable over a larger number of quantum states. Thus, an avalanche results. The *total* energy does not change in this process, but, because of the second law of thermodynamics, avalanches have only been observed in one direction and that is toward the "ground state" — the state with the largest number of ways in which the available energy could be distributed.

Such a collapse (a *decay event*) requires a specific activation energy. For a snow avalanche, this energy comes as a disturbance from outside the system, although such disturbances can be arbitrarily small. In the case of an excited atomic nucleus, the arbitrarily small disturbance comes from quantum vacuum fluctuations. A radioactive nucleus (or any excited system in quantum mechanics) is unstable, and can, thus, *spontaneously* stabilize to a less-excited system. The resulting transformation alters the structure of the nucleus and results in the emission of either a photon or a high-velocity particle that has mass (such as an electron, alpha particle, or other type).

8.9 Occurrence and applications

According to the Big Bang theory, stable isotopes of the lightest five elements (H, He, and traces of Li, Be, and B) were produced very shortly after the emergence of the universe, in a process called Big Bang nucleosynthesis. These lightest stable nuclides (including deuterium) survive to today, but any radioactive isotopes of the light elements produced in the Big Bang (such as tritium) have long since decayed. Isotopes of elements heavier than boron were not produced at all in the Big Bang, and these first five elements do not have any long-lived radioisotopes. Thus, all radioactive nuclei are, therefore, relatively young with respect to the birth of the universe, having formed later in various other types of nucleosynthesis in stars (in particular, supernovae), and also during ongoing interactions between stable isotopes and energetic particles. For example, carbon-14, a radioactive nuclide with a half-life of only 5,730 years, is constantly produced in Earth's upper atmosphere due to interactions between cosmic rays and nitrogen.

Nuclides that are produced by radioactive decay are called radiogenic nuclides, whether they themselves are stable or not. There exist stable radiogenic nuclides that were formed from short-lived extinct radionuclides in the early solar system.[32][33] The extra presence of these stable radiogenic nuclides (such as Xe-129 from primordial I-129) against the background of primordial stable nuclides can be inferred by various means.

Radioactive decay has been put to use in the technique of radioisotopic labeling, which is used to track the passage of a chemical substance through a complex system (such as a living organism). A sample of the substance is synthesized with a high concentration of unstable atoms. The presence of the substance in one or another part of the system is determined by detecting the locations of decay events.

On the premise that radioactive decay is truly random (rather than merely chaotic), it has been used in hardware random-number generators. Because the process is not thought to vary significantly in mechanism over time, it is also a valuable tool in estimating the absolute ages of certain materials. For geological materials, the radioisotopes and some of their decay products become trapped when a rock solidifies, and can then later be used (subject to many well-known qualifications) to estimate the date of the solidification. These include checking the results of several simultaneous processes and their products against each other, within the same sample. In a similar fashion, and also subject to qualification, the rate of formation of carbon-14 in various eras, the date of formation of organic matter within a certain period related to the isotope's half-life may be estimated, because the carbon-14 becomes trapped when the organic matter grows and incorporates the new carbon-14 from the air. Thereafter, the amount of carbon-14 in organic matter decreases according to decay processes that may also be independently cross-checked by other means (such as checking the carbon-14 in individual tree rings, for example).

8.10   Origins of radioactive nuclides

Main article: nucleosynthesis

Radioactive primordial nuclides found in the Earth are residues from ancient supernova explosions which occurred before the formation of the solar system. They are the long-lived fraction of radionuclides surviving in the primordial solar nebula through planet accretion until the present. The naturally occurring short-lived radiogenic radionuclides found in rocks are the daughters of these radioactive primordial nuclides. Another minor source of naturally occurring radioactive nuclides are cosmogenic nuclides, formed by cosmic ray bombardment of material in the Earth's atmosphere or crust. The radioactive decay of these radionuclides in rocks within Earth's mantle and crust contribute significantly to Earth's internal heat budget.

8.11   Decay chains and multiple modes

The daughter nuclide of a decay event may also be unstable (radioactive). In this case, it will also decay, producing radiation. The resulting second daughter nuclide may also be radioactive. This can lead to a sequence of several decay events. Eventually, a stable nuclide is produced. This is called a *decay chain* (see this article for specific details of important natural decay chains).

An example is the natural decay chain of ^{238}U, which is as follows:

- decays, through alpha-emission, with a half-life of 4.5 billion years to thorium-234

- which decays, through beta-emission, with a half-life of 24 days to protactinium-234

- which decays, through beta-emission, with a half-life of 1.2 minutes to uranium-234

- which decays, through alpha-emission, with a half-life of 240 thousand years to thorium-230

- which decays, through alpha-emission, with a half-life of 77 thousand years to radium-226

- which decays, through alpha-emission, with a half-life of 1.6 thousand years to radon-222

- which decays, through alpha-emission, with a half-life of 3.8 days to polonium-218

- which decays, through alpha-emission, with a half-life of 3.1 minutes to lead-214

- which decays, through beta-emission, with a half-life of 27 minutes to bismuth-214

- which decays, through beta-emission, with a half-life of 20 minutes to polonium-214

- which decays, through alpha-emission, with a half-life of 160 microseconds to lead-210

- which decays, through beta-emission, with a half-life of 22 years to bismuth-210

- which decays, through beta-emission, with a half-life of 5 days to polonium-210

- which decays, through alpha-emission, with a half-life of 140 days to lead-206, which is a stable nuclide.

Some radionuclides may have several different paths of decay. For example, approximately 36% of bismuth-212 decays, through alpha-emission, to thallium-208 while approximately 64% of bismuth-212 decays, through beta-emission, to polonium-212. Both thallium-208 and polonium-212 are radioactive daughter products of bismuth-212, and both decay directly to stable lead-208.

8.12 Associated hazard warning signs

- The trefoil symbol used to indicate ionising radiation.

- 2007 ISO radioactivity danger symbol intended for IAEA Category 1, 2 and 3 sources defined as dangerous sources capable of death or serious injury.[1]

- The dangerous goods transport classification sign for radioactive materials

1. ^ IAEA news release Feb 2007

8.13 See also

- Actinides in the environment

- Background radiation

- Chernobyl disaster

- Crimes involving radioactive substances

- Decay chain

- Fallout shelter

- Half-life

- Lists of nuclear disasters and radioactive incidents

- National Council on Radiation Protection and Measurements

- Nuclear engineering

- Nuclear medicine

- Nuclear pharmacy

- Nuclear physics

- Nuclear power

- Particle decay

- Poisson process

- Radiation

- Radiation therapy

- Radioactive contamination

- Radioactivity in biology

- Radiometric dating

- Radionuclide a.k.a. "radio-isotope"

- Secular equilibrium

- Transient equilibrium

8.14   Notes

[1] Radionuclide is the more correct term, but radioisotope is also used. The difference between isotope and nuclide is explained at Isotope#Isotope vs. nuclide.

8.15   References

8.15.1   Inline

[1] "Decay and Half Life". Retrieved 2009-12-14.

[2] Stabin, Michael G. (2007). "3". *Radiation Protection and Dosimetry: An Introduction to Health Physics*. Springer. doi:10.1007/97 0-387-49983-3. ISBN 978-0387499826.

[3] Best, Lara; Rodrigues, George; Velker, Vikram (2013). "1.3". *Radiation Oncology Primer and Review*. Demos Medical Publishing. ISBN 978-1620700044.

[4] Loveland, W.; Morrissey, D.; Seaborg, G.T. (2006). *Modern Nuclear Chemistry*. Wiley-Interscience. p. 57. ISBN 0-471-11532-0.

[5] Mould, Richard F. (1995). *A century of X-rays and radioactivity in medicine : with emphasis on photographic records of the early years* (Reprint. with minor corr ed.). Bristol: Inst. of Physics Publ. p. 12. ISBN 9780750302241.

[6] Kasimir Fajans, "Radioactive transformations and the periodic system of the elements". Berichte der Deutschen Chemischen Gesellschaft, Nr. 46, 1913, p. 422–439

[7] Frederick Soddy, "The Radio Elements and the Periodic Law", Chem. News, Nr. 107, 1913, p.97–99

[8] Sansare, K.; Khanna, V.; Karjodkar, F. (2011). "Early victims of X-rays: a tribute and current perception". *Dentomaxillofacial Radiology* **40** (2): 123–125. doi:10.1259/dmfr/73488299. ISSN 0250-832X. PMC 3520298. PMID 21239576.

[9] Ronald L. Kathern and Paul L. Ziemer, he First Fifty Years of Radiation Protection, physics.isu.edu

[10] Hrabak, M.; Padovan, R. S.; Kralik, M.; Ozretic, D.; Potocki, K. (July 2008). "Nikola Tesla and the Discovery of X-rays". *RadioGraphics* **28** (4): 1189–92. doi:10.1148/rg.284075206. PMID 18635636.

[11] Geoff Meggitt (2008), *Taming the Rays - A history of Radiation and Protection.*, Lulu.com, ISBN 978-1-4092-4667-1

[12] Clarke, R.H.; J. Valentin (2009). "The History of ICRP and the Evolution of its Policies" (PDF). *Annals of the ICRP*. ICRP Publication 109 **39** (1): pp. 75–110. doi:10.1016/j.icrp.2009.07.009. Retrieved 12 May 2012.

[13] Rutherford, Ernest (6 October 1910). "Radium Standards and Nomenclature". *Nature* **84** (2136): 430–431.

[14] *10 CFR 20.1005*. US Nuclear Regulatory Commission. 2009.

[15] The Council of the European Communities (1979-12-21). "Council Directive 80/181/EEC of 20 December 1979 on the approximation of the laws of the Member States relating to Unit of measurement and on the repeal of Directive 71/354/EEC". Retrieved 19 May 2012.

[16] Radioactive Decay

[17] Patel, S.B. (2000). *Nuclear physics : an introduction*. New Delhi: New Age International. pp. 62–72. ISBN 9788122401257.

[18] Introductory Nuclear Physics, K.S. Krane, 1988, John Wiley & Sons Inc, ISBN 978-0-471-80553-3

[19] Cetnar, Jerzy (May 2006). "General solution of Bateman equations for nuclear transmutations". *Annals of Nuclear Energy* **33** (7): 640–645. doi:10.1016/j.anucene.2006.02.004.

[20] K.S. Krane (1988). *Introductory Nuclear Physics*. John Wiley & Sons Inc. p. 164. ISBN 978-0-471-80553-3.

[21] Wang, B.; Yan, S.; Limata, B. et al. (2006). "Change of the 7Be electron capture half-life in metallic environments". *The European Physical Journal A* **28** (3): 375–377. Bibcode:2006EPJA...28..375W. doi:10.1140/epja/i2006-10068-x. ISSN 1434-6001.

[22] Jung, M.; Bosch, F.; Beckert, K. et al. (1992). "First observation of bound-state β⁻ decay". *Physical Review Letters* **69** (15): 2164–2167. Bibcode:1992PhRvL..69.2164J. doi:10.1103/PhysRevLett.69.2164. ISSN 0031-9007. PMID 10046415.

[23] Smoliar, M.I.; Walker, R.J.; Morgan, J.W. (1996). "Re-Os ages of group IIA, IIIA, IVA, and IVB iron meteorites". *Science* **271** (5252): 1099–1102. Bibcode:1996Sci...271.1099S. doi:10.1126/science.271.5252.1099.

[24] Bosch, F.; Faestermann, T.; Friese, J.; Heine, F.; Kienle, P.; Wefers, E.; Zeitelhack, K.; Beckert, K.; Franzke, B.; Klepper, O.; Kozhuharov, C.; Menzel, G.; Moshammer, R.; Nolden, F.; Reich, H.; Schlitt, B.; Steck, M.; Stöhlker, T.; Winkler, T.; Takahashi, K. (1996). "Observation of bound-state β– decay of fully ionized ^{187}Re:^{187}Re-^{187}Os Cosmochronometry". *Physical Review Letters* **77** (26): 5190–5193. Bibcode:1996PhRvL..77.5190B. doi:10.1103/PhysRevLett.77.5190. PMID 10062738.

[25] Emery, G.T. (1972). "Perturbation of Nuclear Decay Rates" (PDF). *Annual Review of Nuclear Science* (ACS Publications) **22**: 165–202. Bibcode:1972ARNPS..22..165E. doi:10.1146/annurev.ns.22.120172.001121. Retrieved 6 August 2012.

[26] "The mystery of varying nuclear decay". *Physics World*. 2 October 2008.

[27] Jenkins, Jere H.; Fischbach, Ephraim (2009). "Perturbation of Nuclear Decay Rates During the Solar Flare of 13 December 2006".*Astroparticle Physics***31**(6):407–411.arXiv:0808.3156.Bibcode:2009APh....31..407J.doi:10.1016/j.astropartphys.2009.

[28] Jenkins, J. H.; Buncher, John B.; Gruenwald, John T.; Krause, Dennis E.; Mattes, Joshua J. et al. (2009). "Evidence of correlations between nuclear decay rates and Earth–Sun distance". *Astroparticle Physics* **32** (1): 42–46. arXiv:0808.3283. Bibcode:2009APh....32...42J. doi:10.1016/j.astropartphys.2009.05.004.

[29] Peter A. Sturrock, Gideon Steinitz, Ephraim Fischbach, Daniel Javorsek, II, Jere H. Jenkins, Analysis of Gamma Radiation from a Radon Source: Indications of a Solar Influence, Accessed on line September 2, 2012.

[30] Norman, E. B.; Shugart, Howard A.; Joshi, Tenzing H.; Firestone, Richard B. et al. (2009). "Evidence against correlations between nuclear decay rates and Earth–Sun distance" (PDF). *Astroparticle Physics* **31** (2): 135–137. arXiv:0810.3265. Bibcode:2009APh....31..135N. doi:10.1016/j.astropartphys.2008.12.004.

[31] NUBASE evaluation of nuclear and decay properties

[32] Clayton, Donald D. (1983). *Principles of Stellar Evolution and Nucleosynthesis* (2nd ed.). University of Chicago Press. p. 75. ISBN 0-226-10953-4.

[33] Bolt, B. A.; Packard, R. E.; Price, P. B. (2007). "John H. Reynolds, Physics: Berkeley". The University of California, Berkeley. Retrieved 2007-10-01.

8.15.2 General

- "Radioactivity", Encyclopædia Britannica. 2006. Encyclopædia Britannica Online. December 18, 2006

- Radio-activity by Ernest Rutherford Phd, Encyclopædia Britannica Eleventh Edition

8.16 External links

- The Lund/LBNL Nuclear Data Search – Contains tabulated information on radioactive decay types and energies.

- Nomenclature of nuclear chemistry

- Specific activity and related topics.

- The Live Chart of Nuclides – IAEA

- Health Physics Society Public Education Website

- Beach, Chandler B., ed. (1914). "Becquerel Rays". *The New Student's Reference Work*. Chicago: F. E. Compton and Co.

- Annotated bibliography for radioactivity from the Alsos Digital Library for Nuclear Issues

- Stochastic Java applet on the decay of radioactive atoms by Wolfgang Bauer

- Stochastic Flash simulation on the decay of radioactive atoms by David M. Harrison

- "Henri Becquerel: The Discovery of Radioactivity", Becquerel's 1896 articles online and analyzed on *BibNum* [click 'à télécharger' for English version].

- "Radioactive change", Rutherford & Soddy article (1903), online and analyzed on *Bibnum* [click 'à télécharger' for English version].

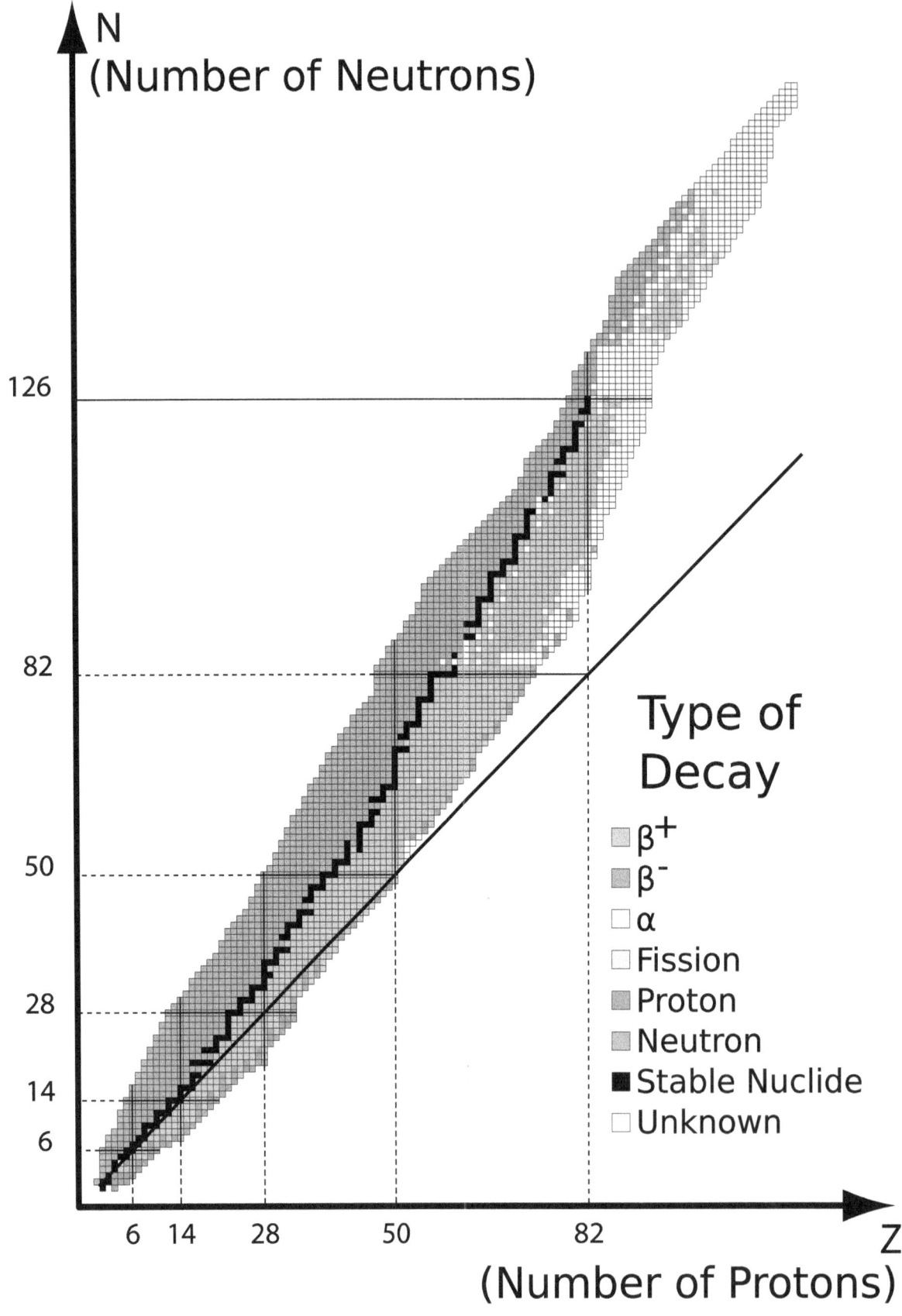

Types of radioactive decay related to N and Z numbers

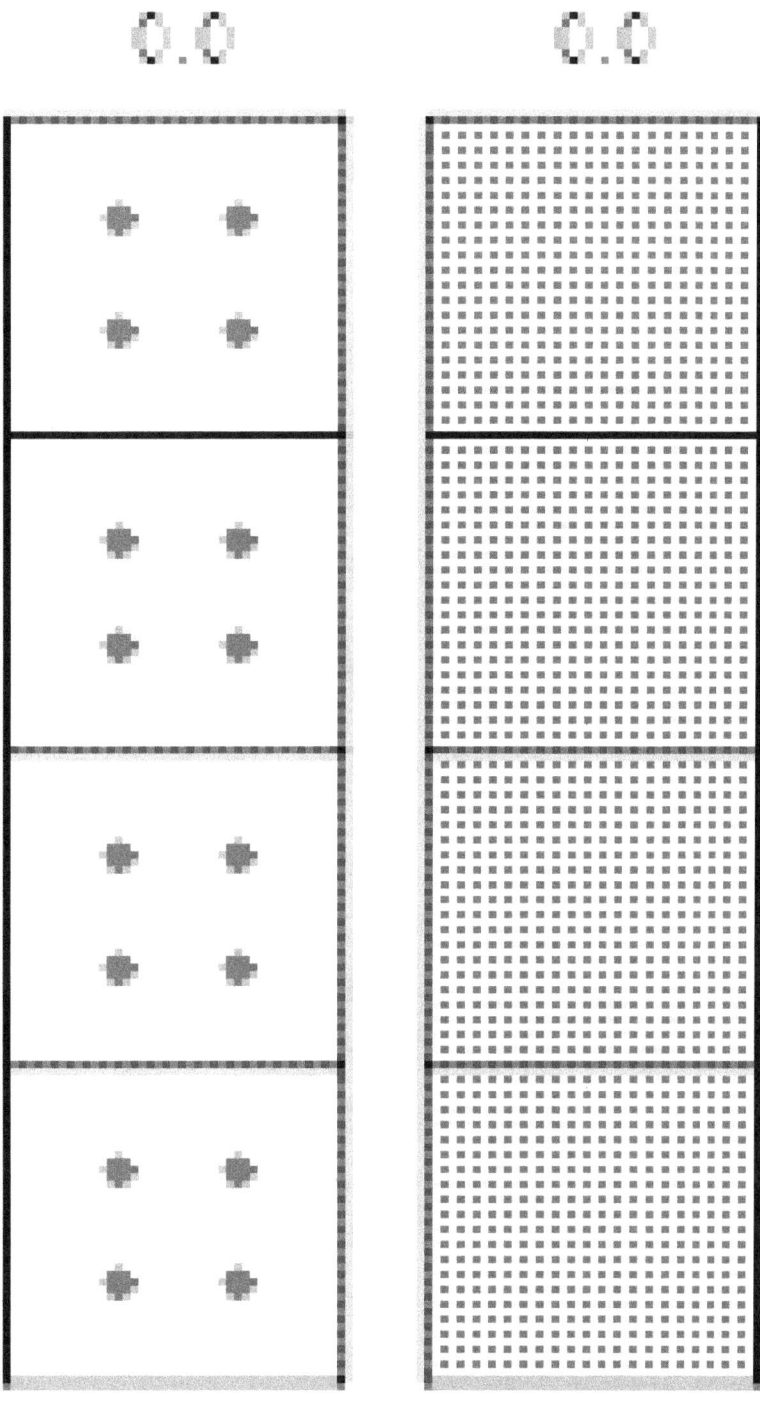

Simulation of many identical atoms undergoing radioactive decay, starting with either 4 atoms (left) or 400 (right). The number at the top indicates how many half-lives have elapsed. Note the law of large numbers: with more atoms, the overall decay is less random.

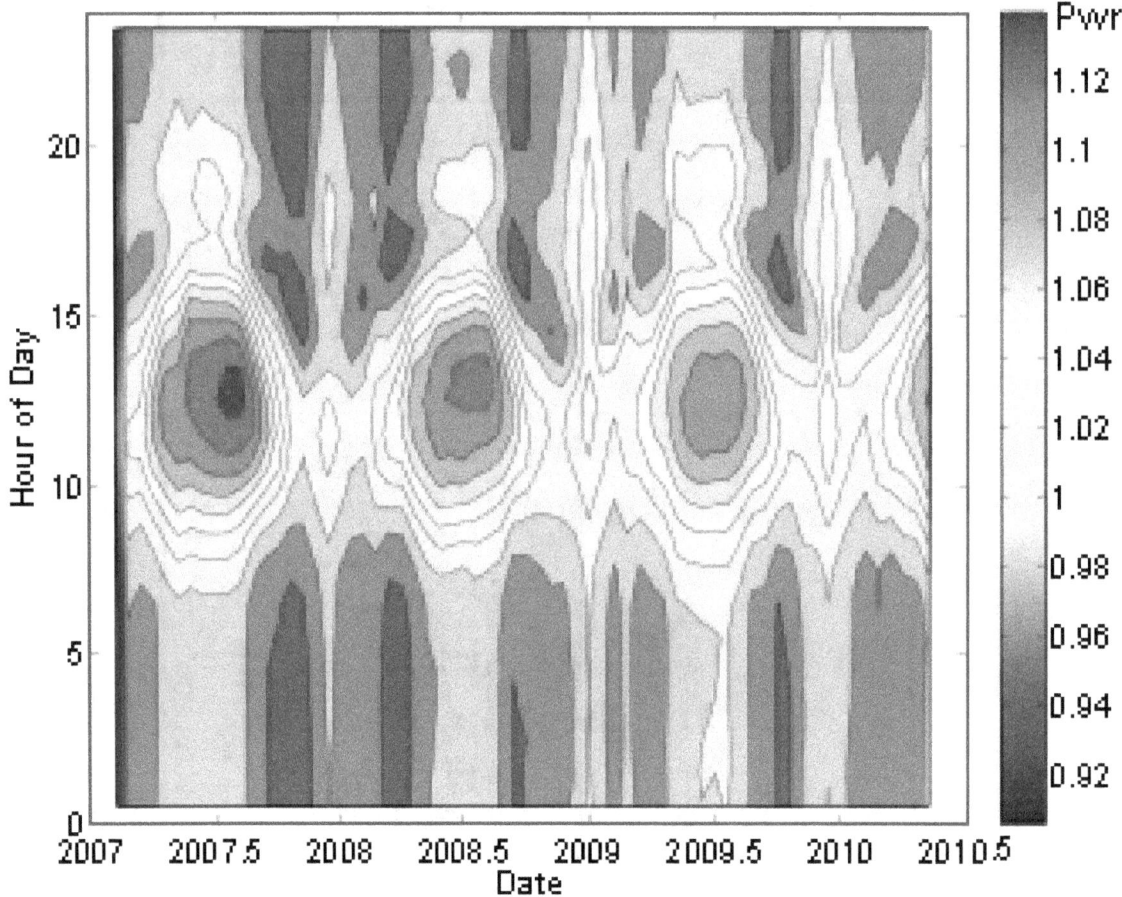

Decay Rate of Radon-222 as a function of date and time of day. The color-bar gives the power of the observed signal and represents ~4% seasonal decay rate variation.

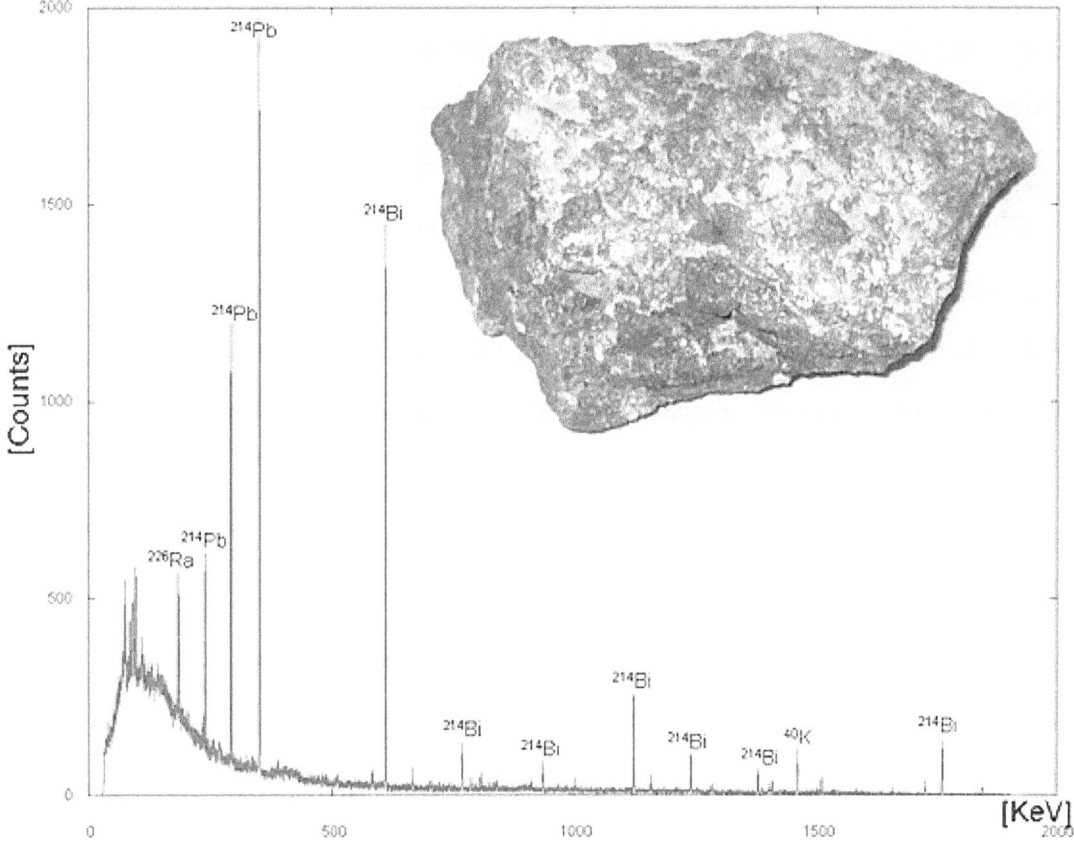

Gamma-ray energy spectrum of uranium ore (inset). Gamma-rays are emitted by decaying nuclides, and the gamma-ray energy can be used to characterize the decay (which nuclide is decaying to which). Here, using the gamma-ray spectrum, several nuclides that are typical of the decay chain of ^{238}U have been identified: ^{226}Ra, ^{214}Pb, ^{214}Bi.

Chapter 9

Electron capture

This article is about the radioactive decay mode. For the fragmentation method used in mass spectrometry, see Electron capture ionization. For the detector used in gas chromatography, see Electron-capture dissociation.

 Electron capture is a process in which the proton-rich nucleus of an electrically neutral atom absorbs an inner atomic electron, thereby changing a nuclear proton to a neutron and simultaneously causing the emission of an electron neutrino. The atom, now in an excited state, then transitions to its ground state. An outer electron replaces the electron that was captured and an X-ray photon is emitted. Electron capture sometimes results in the Auger effect, where an electron is ejected from the atom and a positive ion results. Sometimes, a gamma ray is emitted because the nucleus is also temporarily in an excited state. Following electron capture, the atomic number is reduced by one, but there is no change in atomic mass. Electron capture is an example of weak interaction, one of the four fundamental forces.

Electron capture is the primary decay mode for isotopes with a relative superabundance of protons in the nucleus, but with insufficient energy difference between the isotope and its prospective daughter (the isobar with one less positive charge) for the nuclide to decay by emitting a positron. Electron capture is an alternate decay mode for radioactive isotopes with insufficient energy to decay by positron emission. It is sometimes called **inverse beta decay**, though this term can also refer to the interaction of an electron antineutrino with a proton.[1]

If the energy difference between the parent atom and the daughter atom is less than 1.022 MeV, positron emission is forbidden as not enough decay energy is available to allow it, and thus electron capture is the sole decay mode. For example, rubidium-83 (37 protons, 46 neutrons) will decay to krypton-83 (36 protons, 47 neutrons) solely by electron capture (the energy difference, or decay energy, is about 0.9 MeV).

A free proton cannot normally be changed to a free neutron by this process; the proton and neutron must be part of a larger nucleus. In the process of electron capture, one of the orbital electrons, usually from the K or L electron shell (**K-electron capture**, also **K-capture**, or **L-electron capture**, **L-capture**), is captured by a proton in the nucleus, forming a neutron and emitting an electron neutrino.

Since a proton is changed to a neutron during electron capture, the number of neutrons in the nucleus increases by 1, the number of protons decreases by 1, and the atomic mass number remains unchanged. By changing the number of protons, electron capture transforms the nuclide into a new element. The atom, although still neutral in charge, now exists in an excited state with the inner shell missing an electron. An outer shell electron eventually makes a transition to replace the missing inner electron and thereby moves into a lower energy state. During this process, that electron will emit an X-ray photon (a type of electromagnetic radiation) and other electrons may also be emitted (see Auger electrons). Often the nucleus will be in an excited state also, and will emit a gamma ray as it transitions to the ground state energy of the new nuclide.

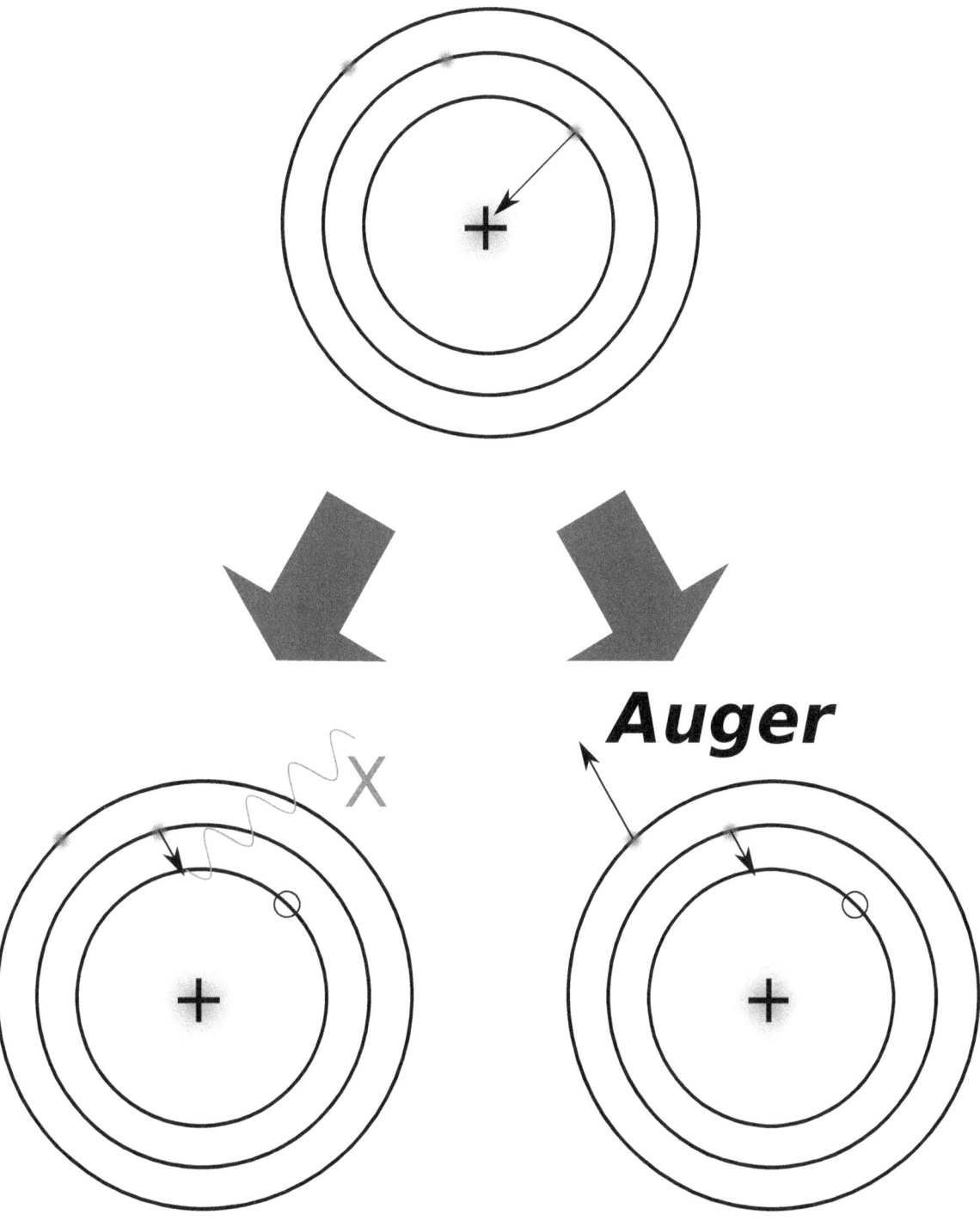

Scheme of two types of electron capture. Top: The nucleus absorbs an electron. Lower left: An outer electron replaces the "missing" electron. An x-ray, equal in energy to the difference between the two electron shells, is emitted. Lower right: In the Auger effect, the energy released when the outer electron replaces the inner electron is transferred to an outer electron. The outer electron is ejected from the atom, leaving a positive ion.

9.1 History

The theory of electron capture was first discussed by Gian-Carlo Wick in a 1934 paper, and then developed by Hideki Yukawa and others. K-electron capture was first observed by Luis Alvarez, in vanadium-48. He reported it in a 1937 paper in *Physical Review*.[2][3][4] Alvarez went on to study electron capture in gallium-67 and other nuclides.[2][5][6]

9.2 Reaction details

The electron that is captured is one of the atom's own electrons, and not a new, incoming electron, as might be suggested by the way the above reactions are written. Radioactive isotopes that decay by pure electron capture can be inhibited from radioactive decay if they are fully ionized ("stripped" is sometimes used to describe such ions). It is hypothesized that such elements, if formed by the r-process in exploding supernovae, are ejected fully ionized and so do not undergo radioactive decay as long as they do not encounter electrons in outer space. Anomalies in elemental distributions are thought to be partly a result of this effect on electron capture. Inverse decays can also be induced by full ionisation; for instance, ^{163}Ho decays into ^{163}Dy by electron capture; however, a fully ionised ^{163}Dy decays into a bound state of ^{163}Ho by the process of bound-state β^- decay.[7]

Chemical bonds can also affect the rate of electron capture to a small degree (in general, less than 1%) depending on the proximity of electrons to the nucleus. For example in ^{7}Be, a difference of 0.9% has been observed between half-lives in metallic and insulating environments.[8] This relatively large effect is due to the fact that beryllium is a small atom whose valence electrons are close to the nucleus.

Around the elements in the middle of the periodic table, isotopes that are lighter than stable isotopes of the same element tend to decay through electron capture, while isotopes heavier than the stable ones decay by electron emission. Electron capture happens most often in the heavier neutron-deficient elements where the mass change is smallest and positron emission isn't always possible. When the loss of mass in a nuclear reaction is greater than zero but less than 2m[0-1e-], the process cannot occur by positron emission but is spontaneous for electron capture.

9.3 Common examples

Some common radioisotopes that decay by electron capture include:

For a full list, see the table of nuclides.

9.4 References

[1] "The Reines-Cowan Experiments: Detecting the Poltergeist" (PDF). *Los Alamos National Laboratory* **25**: 3. 1997.

[2] Luis W. Alvarez, W. Peter Trower (1987). "Chapter 3: K-Electron Capture by Nuclei (with the commentary of Emilio Segré)" In *Discovering Alvarez: selected works of Luis W. Alvarez, with commentary by his students and colleagues.* University of Chicago Press, pp. 11–12, ISBN 978-0-226-81304-2.

[3] "Luis Alvarez, The Nobel Prize in Physics 1968", biography, nobelprize.org. Accessed October 7, 2009.

[4] Alvarez, Luis W. (1937). "Nuclear K Electron Capture". *Physical Review* **52**: 134–135. Bibcode:1937PhRv...52..134A. doi:10.1103/PhysRev.52.134.

[5] Alvarez, Luis W. (1937). "Electron Capture and Internal Conversion in Gallium 67". *Physical Review* **53**:606. Bibcode:1938PhRv...53..A. doi:10.1103/PhysRev.53.606.

[6] Alvarez, Luis W. (1938). "The Capture of Orbital Electrons by Nuclei". *Physical Review* **54**: 486–497. Bibcode:1938PhRv...54..486A. doi:10.1103/PhysRev.54.486.

[7] Fritz Bosch (1995). "Manipulation of Nuclear Lifetimes in Storage Rings" (PDF). *Physica Scripta* **T59**: 221–229.

[8] B. Wang et al. (2006). "Change of the ^{7}Be electron capture half-life in metallic environments". *The European Physical Journal A* **28**: 375–377. (subscription required)

9.5 External links

-  **The LIVEChart of Nuclides - IAEA** with filter on electron capture

Chapter 10

Quantum chromodynamics

In theoretical physics, **quantum chromodynamics (QCD)** is the theory of strong interactions, a fundamental force describing the interactions between quarks and gluons which make up hadrons such as the proton, neutron and pion. QCD is a type of quantum field theory called a non-abelian gauge theory with symmetry group SU(3). The QCD analog of electric charge is a property called *color*. Gluons are the force carrier of the theory, like photons are for the electromagnetic force in quantum electrodynamics. The theory is an important part of the Standard Model of particle physics. A huge body of experimental evidence for QCD has been gathered over the years.

QCD enjoys two peculiar properties:

- **Confinement**, which means that the force between quarks does not diminish as they are separated. Because of this, when you do separate a quark from other quarks, the energy in the gluon field is enough to create another quark pair; they are thus forever bound into hadrons such as the proton and the neutron or the pion and kaon. Although analytically unproven, confinement is widely believed to be true because it explains the consistent failure of free quark searches, and it is easy to demonstrate in lattice QCD.

- **Asymptotic freedom**, which means that in very high-energy reactions, quarks and gluons interact very weakly creating a quark–gluon plasma. This prediction of QCD was first discovered in the early 1970s by David Politzer and by Frank Wilczek and David Gross. For this work they were awarded the 2004 Nobel Prize in Physics.

The phase transition temperature between these two properties has been measured by the ALICE experiment to be well above 160 MeV.[1] Below this temperature, confinement is dominant, while above it, asymptotic freedom becomes dominant.

10.1   Terminology

The word *quark* was coined by American physicist Murray Gell-Mann (b. 1929) in its present sense. It originally comes from the phrase "Three quarks for Muster Mark" in *Finnegans Wake* by James Joyce. On June 27, 1978, Gell-Mann wrote a private letter to the editor of the *Oxford English Dictionary*, in which he related that he had been influenced by Joyce's words: "The allusion to three quarks seemed perfect." (Originally, only three quarks had been discovered.) Gell-Mann, however, wanted to pronounce the word to rhyme with "fork" rather than with "park", as Joyce seemed to indicate by rhyming words in the vicinity such as *Mark*. Gell-Mann got around that "by supposing that one ingredient of the line 'Three quarks for Muster Mark' was a cry of 'Three quarts for Mister ...' heard in H.C. Earwicker's pub", a plausible suggestion given the complex punning in Joyce's novel.[2]

The three kinds of charge in QCD (as opposed to one in quantum electrodynamics or QED) are usually referred to as "color charge" by loose analogy to the three kinds of color (red, green and blue) perceived by humans. Other than this nomenclature, the quantum parameter "color" is completely unrelated to the everyday, familiar phenomenon of color.

Since the theory of electric charge is dubbed "electrodynamics", the Greek word "chroma" Χρώμα (meaning color) is applied to the theory of color charge, "chromodynamics".

10.2   History

With the invention of bubble chambers and spark chambers in the 1950s, experimental particle physics discovered a large and ever-growing number of particles called hadrons. It seemed that such a large number of particles could not all be fundamental. First, the particles were classified by charge and isospin by Eugene Wigner and Werner Heisenberg; then, in 1953, according to strangeness by Murray Gell-Mann and Kazuhiko Nishijima. To gain greater insight, the hadrons were sorted into groups having similar properties and masses using the *eightfold way*, invented in 1961 by Gell-Mann and Yuval Ne'eman. Gell-Mann and George Zweig, correcting an earlier approach of Shoichi Sakata, went on to propose in 1963 that the structure of the groups could be explained by the existence of three flavors of smaller particles inside the hadrons: the quarks.

Perhaps the first remark that quarks should possess an additional quantum number was made[3] as a short footnote in the preprint of Boris Struminsky[4] in connection with Ω^- hyperon composed of three strange quarks with parallel spins (this situation was peculiar, because since quarks are fermions, such combination is forbidden by the Pauli exclusion principle):

> Three identical quarks cannot form an antisymmetric S-state. In order to realize an antisymmetric orbital S-state, it is necessary for the quark to have an additional quantum number.
> — B. V. Struminsky, *Magnetic moments of barions in the quark model*, JINR-Preprint P-1939, Dubna, Submitted on January 7, 1965

Boris Struminsky was a PhD student of Nikolay Bogolyubov. The problem considered in this preprint was suggested by Nikolay Bogolyubov, who advised Boris Struminsky in this research.[4] In the beginning of 1965, Nikolay Bogolyubov, Boris Struminsky and Albert Tavkhelidze wrote a preprint with a more detailed discussion of the additional quark quantum degree of freedom.[5] This work was also presented by Albert Tavchelidze without obtaining consent of his collaborators for doing so at an international conference in Trieste (Italy), in May 1965.[6][7]

A similar mysterious situation was with the Δ^{++} baryon; in the quark model, it is composed of three up quarks with parallel spins. In 1965, Moo-Young Han with Yoichiro Nambu and Oscar W. Greenberg independently resolved the problem by proposing that quarks possess an additional SU(3) gauge degree of freedom, later called color charge. Han and Nambu noted that quarks might interact via an octet of vector gauge bosons: the gluons.

Since free quark searches consistently failed to turn up any evidence for the new particles, and because an elementary particle back then was *defined* as a particle which could be separated and isolated, Gell-Mann often said that quarks were merely convenient mathematical constructs, not real particles. The meaning of this statement was usually clear in context: He meant quarks are confined, but he also was implying that the strong interactions could probably not be fully described by quantum field theory.

Richard Feynman argued that high energy experiments showed quarks are real particles: he called them *partons* (since they were parts of hadrons). By particles, Feynman meant objects which travel along paths, elementary particles in a field theory.

The difference between Feynman's and Gell-Mann's approaches reflected a deep split in the theoretical physics community. Feynman thought the quarks have a distribution of position or momentum, like any other particle, and he (correctly) believed that the diffusion of parton momentum explained diffractive scattering. Although Gell-Mann believed that certain quark charges could be localized, he was open to the possibility that the quarks themselves could not be localized because space and time break down. This was the more radical approach of S-matrix theory.

James Bjorken proposed that pointlike partons would imply certain relations should hold in deep inelastic scattering of electrons and protons, which were spectacularly verified in experiments at SLAC in 1969. This led physicists to abandon the S-matrix approach for the strong interactions.

The discovery of asymptotic freedom in the strong interactions by David Gross, David Politzer and Frank Wilczek allowed physicists to make precise predictions of the results of many high energy experiments using the quantum field theory technique of perturbation theory. Evidence of gluons was discovered in three-jet events at PETRA in 1979. These experiments became more and more precise, culminating in the verification of perturbative QCD at the level of a few percent at the LEP in CERN.

The other side of asymptotic freedom is confinement. Since the force between color charges does not decrease with distance, it is believed that quarks and gluons can never be liberated from hadrons. This aspect of the theory is verified within lattice QCD computations, but is not mathematically proven. One of the Millennium Prize Problems announced by the Clay Mathematics Institute requires a claimant to produce such a proof. Other aspects of non-perturbative QCD are the exploration of phases of quark matter, including the quark–gluon plasma.

The relation between the short-distance particle limit and the confining long-distance limit is one of the topics recently explored using string theory, the modern form of S-matrix theory.[8][9]

10.3 Theory

10.3.1 Some definitions

Every field theory of particle physics is based on certain symmetries of nature whose existence is deduced from observations. These can be

- local symmetries, that is the symmetry acts independently at each point in spacetime. Each such symmetry is the basis of a gauge theory and requires the introduction of its own gauge bosons.

- global symmetries, which are symmetries whose operations must be simultaneously applied to all points of spacetime.

QCD is a gauge theory of the SU(3) gauge group obtained by taking the color charge to define a local symmetry.

Since the strong interaction does not discriminate between different flavors of quark, QCD has approximate **flavor symmetry**, which is broken by the differing masses of the quarks.

There are additional global symmetries whose definitions require the notion of chirality, discrimination between left and right-handed. If the spin of a particle has a positive projection on its direction of motion then it is called left-handed; otherwise, it is right-handed. Chirality and handedness are not the same, but become approximately equivalent at high energies.

- **Chiral** symmetries involve independent transformations of these two types of particle.

- **Vector** symmetries (also called diagonal symmetries) mean the same transformation is applied on the two chiralities.

- **Axial** symmetries are those in which one transformation is applied on left-handed particles and the inverse on the right-handed particles.

10.3.2 Additional remarks: duality

As mentioned, *asymptotic freedom* means that at large energy – this corresponds also to *short distances* – there is practically no interaction between the particles. This is in contrast – more precisely one would say *dual* – to what one is used to, since usually one connects the absence of interactions with *large* distances. However, as already mentioned in the original paper of Franz Wegner,[10] a solid state theorist who introduced 1971 simple gauge invariant lattice models, the high-temperature behaviour of the *original model*, e.g. the strong decay of correlations at large distances, corresponds to the low-temperature behaviour of the (usually ordered!) *dual model*, namely the asymptotic decay of non-trivial correlations, e.g. short-range deviations from almost perfect arrangements, for short distances. Here, in contrast to Wegner, we have only the dual model, which is that one described in this article.[11]

10.3.3 Symmetry groups

The color group SU(3) corresponds to the local symmetry whose gauging gives rise to QCD. The electric charge labels a representation of the local symmetry group U(1) which is gauged to give QED: this is an abelian group. If one considers a version of QCD with *Nf* flavors of massless quarks, then there is a global (chiral) flavor symmetry group SUL(*Nf*) × SUR(*Nf*) × UB(1) × UA(1). The chiral symmetry is spontaneously broken by the QCD vacuum to the vector (L+R) SUV(*Nf*) with the formation of a chiral condensate. The vector symmetry, UB(1) corresponds to the baryon number of quarks and is an exact symmetry. The axial symmetry UA(1) is exact in the classical theory, but broken in the quantum theory, an occurrence called an anomaly. Gluon field configurations called instantons are closely related to this anomaly.

There are two different types of SU(3) symmetry: there is the symmetry that acts on the different colors of quarks, and this is an exact gauge symmetry mediated by the gluons, and there is also a flavor symmetry which rotates different flavors of quarks to each other, or *flavor SU(3)*. Flavor SU(3) is an approximate symmetry of the vacuum of QCD, and is not a fundamental symmetry at all. It is an accidental consequence of the small mass of the three lightest quarks.

In the QCD vacuum there are vacuum condensates of all the quarks whose mass is less than the QCD scale. This includes the up and down quarks, and to a lesser extent the strange quark, but not any of the others. The vacuum is symmetric under SU(2) isospin rotations of up and down, and to a lesser extent under rotations of up, down and strange, or full flavor group SU(3), and the observed particles make isospin and SU(3) multiplets.

The approximate flavor symmetries do have associated gauge bosons, observed particles like the rho and the omega, but these particles are nothing like the gluons and they are not massless. They are emergent gauge bosons in an approximate string description of QCD.

10.3.4 Lagrangian

The dynamics of the quarks and gluons are controlled by the quantum chromodynamics Lagrangian. The gauge invariant QCD Lagrangian is

where $\psi_i(x)$ is the quark field, a dynamical function of spacetime, in the fundamental representation of the SU(3) gauge group, indexed by i, j, ... ; $\mathcal{A}_\mu^a(x)$ are the gluon fields, also dynamical functions of spacetime, in the adjoint representation of the SU(3) gauge group, indexed by a, b,... The γ^μ are Dirac matrices connecting the spinor representation to the vector representation of the Lorentz group.

The symbol $G_{\mu\nu}^a$ represents the gauge invariant gluon field strength tensor, analogous to the electromagnetic field strength tensor, $F^{\mu\nu}$, in quantum electrodynamics. It is given by:[12]

$$G_{\mu\nu}^a = \partial_\mu \mathcal{A}_\nu^a - \partial_\nu \mathcal{A}_\mu^a + g f^{abc} \mathcal{A}_\mu^b \mathcal{A}_\nu^c \,,$$

where *fabc* are the structure constants of SU(3). Note that the rules to move-up or pull-down the a, b, or c indexes are *trivial*, (+, ..., +), so that $f^{abc} = fabc = f^a bc$ whereas for the μ or ν indexes one has the non-trivial *relativistic* rules, corresponding e.g. to the metric signature (+ − − −).

The constants *m* and *g* control the quark mass and coupling constants of the theory, subject to renormalization in the full quantum theory.

An important theoretical notion concerning the final term of the above Lagrangian is the *Wilson loop* variable. This loop variable plays a most important role in discretized forms of the QCD (see lattice QCD), and more generally, it distinguishes confined and deconfined states of a gauge theory. It was introduced by the Nobel prize winner Kenneth G. Wilson and is treated in a separate article.

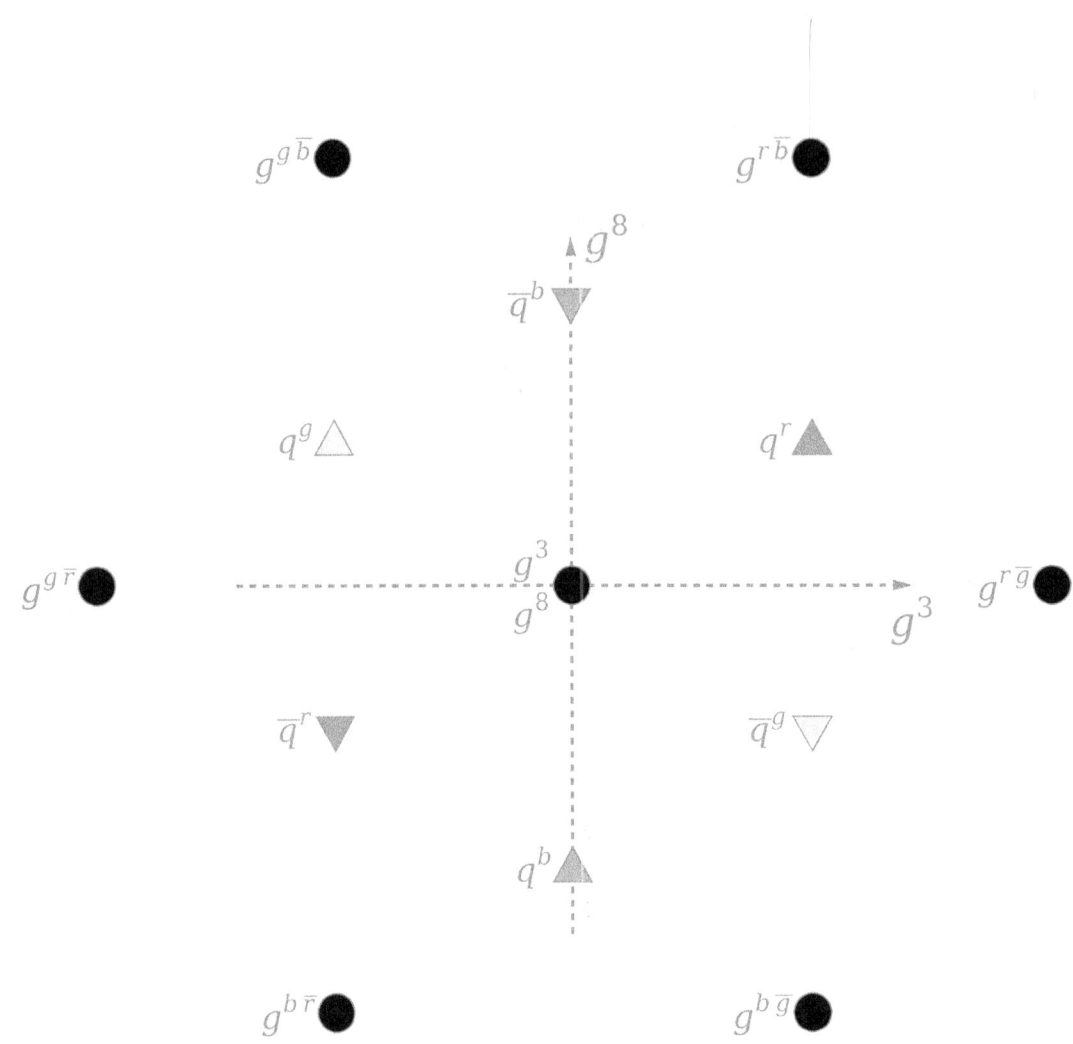

The pattern of strong charges for the three colors of quark, three antiquarks, and eight gluons (with two of zero charge overlapping).

10.3.5 Fields

Quarks are massive spin-1/2 fermions which carry a color charge whose gauging is the content of QCD. Quarks are represented by Dirac fields in the fundamental representation **3** of the gauge group SU(3). They also carry electric charge (either $-1/3$ or $2/3$) and participate in weak interactions as part of weak isospin doublets. They carry global quantum numbers including the baryon number, which is 1/3 for each quark, hypercharge and one of the flavor quantum numbers.

Gluons are spin-1 bosons which also carry color charges, since they lie in the adjoint representation **8** of SU(3). They have no electric charge, do not participate in the weak interactions, and have no flavor. They lie in the singlet representation **1** of all these symmetry groups.

Every quark has its own antiquark. The charge of each antiquark is exactly the opposite of the corresponding quark.

10.3.6   Dynamics

According to the rules of quantum field theory, and the associated Feynman diagrams, the above theory gives rise to three basic interactions: a quark may emit (or absorb) a gluon, a gluon may emit (or absorb) a gluon, and two gluons may directly interact. This contrasts with QED, in which only the first kind of interaction occurs, since photons have no charge. Diagrams involving Faddeev–Popov ghosts must be considered too (except in the unitarity gauge).

10.3.7   Area law and confinement

Detailed computations with the above-mentioned Lagrangian[13] show that the effective potential between a quark and its anti-quark in a meson contains a term $\propto r$, which represents some kind of "stiffness" of the interaction between the particle and its anti-particle at large distances, similar to the entropic elasticity of a rubber band (see below). This leads to *confinement* [14] of the quarks to the interior of hadrons, i.e. mesons and nucleons, with typical radii R_c, corresponding to former "Bag models" of the hadrons[15] . The order of magnitude of the "bag radius" is 1 fm (= 10^{-15} m). Moreover, the above-mentioned stiffness is quantitatively related to the so-called "area law" behaviour of the expectation value of the Wilson loop product PW of the ordered coupling constants around a closed loop W; i.e. $\langle P_W \rangle$ is proportional to the *area* enclosed by the loop. For this behaviour the non-abelian behaviour of the gauge group is essential.

10.4   Methods

Further analysis of the content of the theory is complicated. Various techniques have been developed to work with QCD. Some of them are discussed briefly below.

10.4.1   Perturbative QCD

Main article: Perturbative QCD

This approach is based on asymptotic freedom, which allows perturbation theory to be used accurately in experiments performed at very high energies. Although limited in scope, this approach has resulted in the most precise tests of QCD to date.

10.4.2   Lattice QCD

Main article: Lattice QCD

Among non-perturbative approaches to QCD, the most well established one is lattice QCD. This approach uses a discrete set of spacetime points (called the lattice) to reduce the analytically intractable path integrals of the continuum theory to a very difficult numerical computation which is then carried out on supercomputers like the QCDOC which was constructed for precisely this purpose. While it is a slow and resource-intensive approach, it has wide applicability, giving insight into parts of the theory inaccessible by other means, in particular into the explicit forces acting between quarks and antiquarks in a meson. However, the numerical sign problem makes it difficult to use lattice methods to study QCD at high density and low temperature (e.g. nuclear matter or the interior of neutron stars).

10.4.3   1/N expansion

Main article: 1/N expansion

A well-known approximation scheme, the 1/N expansion, starts from the premise that the number of colors is infinite, and makes a series of corrections to account for the fact that it is not. Until now, it has been the source of qualitative insight rather than a method for quantitative predictions. Modern variants include the AdS/CFT approach.

10.4.4 Effective theories

For specific problems effective theories may be written down which give qualitatively correct results in certain limits. In the best of cases, these may then be obtained as systematic expansions in some parameter of the QCD Lagrangian. One such effective field theory is chiral perturbation theory or ChiPT, which is the QCD effective theory at low energies. More precisely, it is a low energy expansion based on the spontaneous chiral symmetry breaking of QCD, which is an exact symmetry when quark masses are equal to zero, but for the u,d and s quark, which have small mass, it is still a good approximate symmetry. Depending on the number of quarks which are treated as light, one uses either SU(2) ChiPT or SU(3) ChiPT . Other effective theories are heavy quark effective theory (which expands around heavy quark mass near infinity), and soft-collinear effective theory (which expands around large ratios of energy scales). In addition to effective theories, models like the Nambu–Jona-Lasinio model and the chiral model are often used when discussing general features.

10.4.5 QCD sum rules

Main article: QCD sum rules

Based on an Operator product expansion one can derive sets of relations that connect different observables with each other.

10.4.6 Nambu–Jona-Lasinio model

In one of his recent works, Kei-Ichi Kondo derived as a low-energy limit of QCD, a theory linked to the Nambu–Jona-Lasinio model since it is basically a particular non-local version of the Polyakov–Nambu–Jona-Lasinio model.[17] The later being in its local version, nothing but the Nambu–Jona-Lasinio model in which one has included the Polyakov loop effect, in order to describe a 'certain confinement'.

The Nambu–Jona-Lasinio model in itself is, among many other things, used because it is a 'relatively simple' model of chiral symmetry breaking, phenomenon present up to certain conditions (Chiral limit i.e. massless fermions) in QCD itself. In this model, however, there is no confinement. In particular, the energy of an isolated quark in the physical vacuum turns out well defined and finite.

10.5 Experimental tests

The notion of quark flavors was prompted by the necessity of explaining the properties of hadrons during the development of the quark model. The notion of color was necessitated by the puzzle of the $\Delta++$. This has been dealt with in the section on the history of QCD.

The first evidence for quarks as real constituent elements of hadrons was obtained in deep inelastic scattering experiments at SLAC. The first evidence for gluons came in three jet events at PETRA.

Several good quantitative tests of perturbative QCD exist:

- The running of the QCD coupling as deduced from many observations
- Scaling violation in polarized and unpolarized deep inelastic scattering
- Vector boson production at colliders (this includes the Drell-Yan process)
- Jet cross sections in colliders
- Event shape observables at the LEP
- Heavy-quark production in colliders

Quantitative tests of non-perturbative QCD are fewer, because the predictions are harder to make. The best is probably the running of the QCD coupling as probed through lattice computations of heavy-quarkonium spectra. There is a recent claim about the mass of the heavy meson B_c. Other non-perturbative tests are currently at the level of 5% at best. Continuing work on masses and form factors of hadrons and their weak matrix elements are promising candidates for future quantitative tests. The whole subject of quark matter and the quark–gluon plasma is a non-perturbative test bed for QCD which still remains to be properly exploited.

One qualitative prediction of QCD is that there exist composite particles made solely of gluons called glueballs that have not yet been definitively observed experimentally. A definitive observation of a glueball with the properties predicted by QCD would strongly confirm the theory. In principle, if glueballs could be definitively ruled out, this would be a serious experimental blow to QCD. But, as of 2013, scientists are unable to confirm or deny the existence of glueballs definitively, despite the fact that particle accelerators have sufficient energy to generate them.

10.6   Cross-relations to solid state physics

There are unexpected cross-relations to solid state physics. For example, the notion of gauge invariance forms the basis of the well-known Mattis spin glasses,[18] which are systems with the usual spin degrees of freedom $s_i = \pm 1$ for i =1,...,N, with the special fixed "random" couplings $J_{i,k} = \epsilon_i J_0 \epsilon_k$. Here the ϵ_i and ϵ_k quantities can independently and "randomly" take the values ± 1, which corresponds to a most-simple gauge transformation ($s_i \to s_i \cdot \epsilon_i \quad J_{i,k} \to \epsilon_i J_{i,k} \epsilon_k \quad s_k \to s_k \cdot \epsilon_k$). This means that thermodynamic expectation values of measurable quantities, e.g. of the energy $\mathcal{H} := -\sum s_i J_{i,k} s_k$, are invariant.

However, here the *coupling degrees of freedom* $J_{i,k}$, which in the QCD correspond to the *gluons*, are "frozen" to fixed values (quenching). In contrast, in the QCD they "fluctuate" (annealing), and through the large number of gauge degrees of freedom the entropy plays an important role (see below).

For positive J_0 the thermodynamics of the Mattis spin glass corresponds in fact simply to a "ferromagnet in disguise", just because these systems have no "frustration" at all. This term is a basic measure in spin glass theory.[19] Quantitatively it is identical with the loop product $P_W := J_{i,k} J_{k,l} ... J_{n,m} J_{m,i}$ along a closed loop W. However, for a Mattis spin glass – in contrast to "genuine" spin glasses – the quantity PW never becomes negative.

The basic notion "frustration" of the spin-glass is actually similar to the Wilson loop quantity of the QCD. The only difference is again that in the QCD one is dealing with SU(3) matrices, and that one is dealing with a "fluctuating" quantity. Energetically, perfect absence of frustration should be non-favorable and atypical for a spin glass, which means that one should add the loop product to the Hamiltonian, by some kind of term representing a "punishment". In the QCD the Wilson loop is essential for the Lagrangian rightaway.

The relation between the QCD and "disordered magnetic systems" (the spin glasses belong to them) were additionally stressed in a paper by Fradkin, Huberman und Shenker,[20] which also stresses the notion of duality.

A further analogy consists in the already mentioned similarity to polymer physics, where, analogously to Wilson Loops, so-called "entangled nets" appear, which are important for the formation of the entropy-elasticity (force proportional to the length) of a rubber band. The non-abelian character of the SU(3) corresponds thereby to the non-trivial "chemical links", which glue different loop segments together, and "asymptotic freedom" means in the polymer analogy simply the fact that in the short-wave limit, i.e. for $0 \leftarrow \lambda_w \ll R_c$ (where Rc is a characteristic correlation length for the glued loops, corresponding to the above-mentioned "bag radius", while λ_w is the wavelength of an excitation) any non-trivial correlation vanishes totally, as if the system had crystallized.[21]

There is also a correspondence between confinement in QCD – the fact that the color field is only different from zero in the interior of hadrons – and the behaviour of the usual magnetic field in the theory of type-II superconductors: there the magnetism is confined to the interiour of the Abrikosov flux-line lattice,[22] i.e., the London penetration depth λ of that theory is analogous to the confinement radius Rc of quantum chromodynamics. Mathematically, this correspondendence is supported by the second term, $\propto g G_\mu^a \bar{\psi}_i \gamma^\mu T_{ij}^a \psi_j$, on the r.h.s. of the Lagrangian.

10.7 See also

- For overviews, see Standard Model, its field theoretical formulation, strong interactions, quarks and gluons, hadrons, confinement, QCD matter, or quark–gluon plasma.

- For details, see gauge theory, quantization procedure including BRST quantization and Faddeev–Popov ghosts. A more general category is quantum field theory.

- For techniques, see Lattice QCD, 1/N expansion, perturbative QCD, Soft-collinear effective theory, heavy quark effective theory, chiral models, and the Nambu and Jona-Lasinio model.

- For experiments, see quark search experiments, deep inelastic scattering, jet physics, quark–gluon plasma.

- Symmetry in quantum mechanics

10.8 References

[1] "Alice Physics". 26 August 2015.

[2] Gell-Mann, Murray (1995). *The Quark and the Jaguar*. Owl Books. ISBN 978-0-8050-7253-2.

[3] Fyodor Tkachov (2009). "A contribution to the history of quarks: Boris Struminsky's 1965 JINR publication". arXiv:0904.0343 [physics.hist-ph].

[4] B. V. Struminsky, Magnetic moments of barions in the quark model. JINR-Preprint P-1939, Dubna, Russia. Submitted on January 7, 1965.

[5] N. Bogolubov, B. Struminsky, A. Tavkhelidze. On composite models in the theory of elementary particles. JINR Preprint D-1968, Dubna 1965.

[6] A. Tavkhelidze. Proc. Seminar on High Energy Physics and Elementary Particles, Trieste, 1965, Vienna IAEA, 1965, p. 763.

[7] V. A. Matveev and A. N. Tavkhelidze (INR, RAS, Moscow) The quantum number color, colored quarks and QCD (Dedicated to the 40th Anniversary of the Discovery of the Quantum Number Color). Report presented at the 99th Session of the JINR Scientific Council, Dubna, 19–20 January 2006.

[8] J. Polchinski, M. Strassler (2002). "Hard Scattering and Gauge/String duality". *Physical Review Letters* **88** (3): 31601. arXiv:hep-th/0109174. Bibcode:2002PhRvL..88c1601P. doi:10.1103/PhysRevLett.88.031601. PMID 11801052.

[9] Brower, Richard C.; Mathur, Samir D.; Chung-I Tan (2000). "Glueball Spectrum for QCD from AdS Supergravity Duality". *Nuclear Physics* B**587**:249–276.arXiv:hep-th/0003115.Bibcode:2000NuPhB.587..249B.doi:10.1016/S0550-3213(00)0-1.

[10] F. Wegner, *Duality in Generalized Ising Models and Phase Transitions without Local Order Parameter*, J. Math. Phys. **12** (1971) 2259–2272.

Reprinted in Claudio Rebbi (ed.), *Lattice Gauge Theories and Monte Carlo Simulations*, World Scientific, Singapore (1983), p. 60–73. Abstract:

[11] Perhaps one can guess that in the "original" model mainly the quarks would fluctuate, whereas in the present one, the "dual" model, mainly the gluons do.

[12] M. Eidemüller, H.G. Dosch, M. Jamin (1999). "The field strength correlator from QCD sum rules". *Nucl.Phys.Proc.Suppl.86:421–425,2000* (Heidelberg, Germany). arXiv:hep-ph/9908318.

[13] See all standard textbooks on the QCD, e.g., those noted above

[14] Only at extremely large pressures and or temperatures, e.g. for $T \cong 5 \cdot 10^{12}$ K or larger, *confinement* gives way to a quark–gluon plasma.

[15] Kenneth A. Johnson, "The bag model of quark confinement", Scientific American, July 1979

[16] M. Cardoso et al., "Lattice QCD computation of the colour fields for the static hybrid quark–gluon–antiquark system, and microscopic study of the Casimir scaling", Phys. Rev. D 81, 034504 (2010)).

[17] Kei-Ichi Kondo (2010). "Toward a first-principle derivation of confinement and chiral-symmetry-breaking crossover transitions in QCD". *Physical Review D***82**(6):065024.arXiv:1005.0314v2.Bibcode:2010PhRvD..82f5024K.doi:10.1103/PhysRevD.82..

[18] D.C. Mattis, Phys. Lett. 56a (1976) 421

[19] J. Vanninemus and G. Toulouse, J. Phys. C 10 (1977) 537

[20] E. Fradkin, B.A. Huberman, S. Shenker, *Gauge Symmetries in random magnetic systems*, Phys. Rev. B 18 (1978) 4783–4794,

[21] A. Bergmann, A. Owen, "Dielectric relaxation spectroscopy of poly[(R)−3-Hydroxybutyrate] (PHD) during crystallization", Polymer International 53 (7) (2004) 863–868,

[22] Mathematically, the flux-line lattices are described by Emil Artin's braid group, which is nonabelian, since one braid can wind around another one.

10.9   Further reading

- Greiner, Walter;Schäfer, Andreas (1994). *Quantum Chromodynamics*. Springer. ISBN 0-387-57103-5.

- Halzen, Francis; Martin, Alan (1984). *Quarks & Leptons: An Introductory Course in Modern Particle Physics*. John Wiley & Sons. ISBN 0-471-88741-2.

- Creutz, Michael (1985). *Quarks, Gluons and Lattices*. Cambridge University Press. ISBN 978-0-521-31535-7.

10.10   External links

- Frank Wilczek (2000). "QCD made simple" (PDF). *Physics Today* **53** (8): 22–28. doi:10.1063/1.1310117.

- Particle data group

- The millennium prize for proving confinement

- Ab Initio Determination of Light Hadron Masses

- Andreas S Kronfeld *The Weight of the World Is Quantum Chromodynamics*

- Andreas S Kronfeld *Quantum chromodynamics with advanced computing*

- Standard model gets right answer

- Quantum Chromodynamics

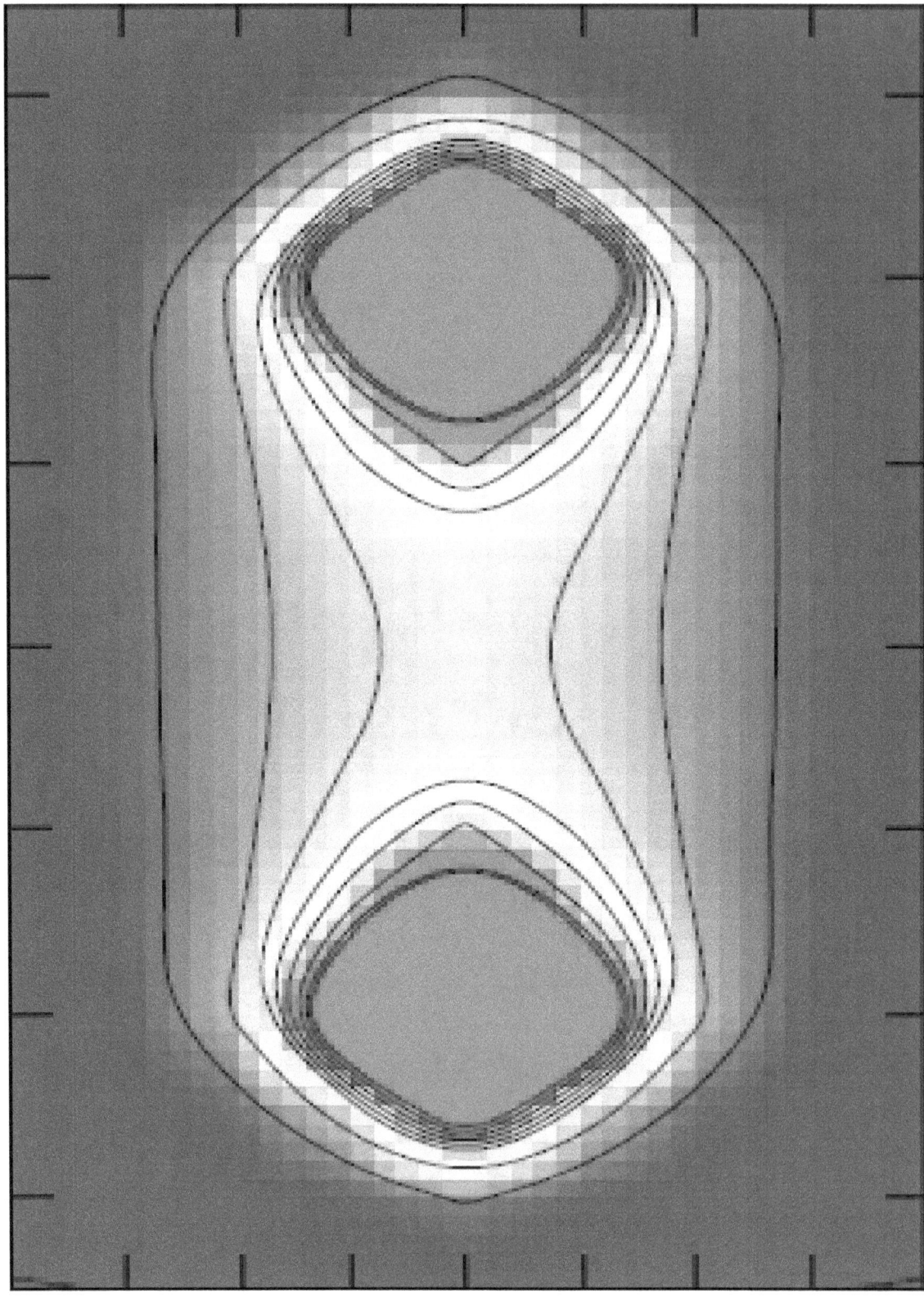

A quark and an antiquark (red color) are glued together (green color) to form a meson (result of a lattice QCD simulation by M. Cardoso et al.

Chapter 11

Proton therapy

Proton therapy equipment at Mayo Clinic in Rochester, Minnesota

Proton therapy or **proton beam therapy** is a medical procedure, a type of particle therapy that uses a beam of protons to irradiate diseased tissue, most often in the treatment of cancer. Proton therapy's chief advantage over other types of external beam radiotherapy is that as a charged particle the dose is deposited over a narrow range and there is minimal exit dose.

11.1 Description

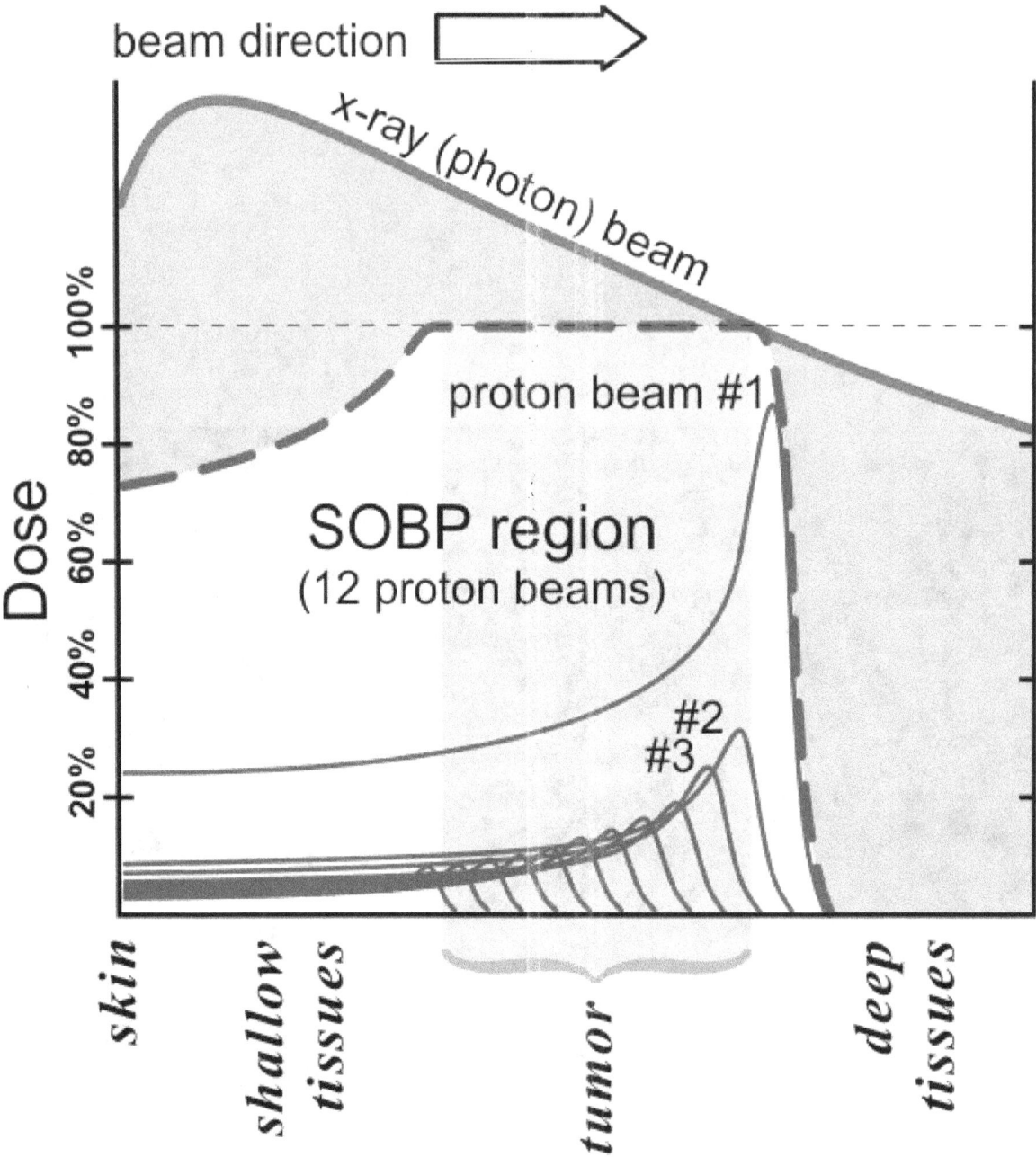

In a typical treatment plan for proton therapy, the spread out bragg peak (SOBP, *dashed blue line) is the therapeutic radiation distribution. The SOBP is the sum of several individual Bragg peaks (thin blue lines) at staggered depths. The depth-dose plot of an X-ray beam (red line) is provided for comparison. The pink area represents additional doses of X-ray radiotherapy—which can damage normal tissues and cause secondary cancers, especially of the skin.[1]*

Main article: Radiation therapy § Mechanism of action

Proton therapy is a type of external beam radiotherapy that uses ionizing radiation. In proton therapy, medical personnel use a particle accelerator to target a tumor with a beam of protons.[2][3] These charged particles damage the DNA of cells, ultimately killing them or stopping their reproduction. Cancerous cells are particularly vulnerable to attacks on DNA because of their high rate of division and their reduced abilities to repair DNA damage.

Due to their relatively large mass, protons have little lateral side scatter in the tissue; the beam does not broaden much, stays focused on the tumor shape and delivers only low-dose side-effects to surrounding tissue. All protons of a given energy have a certain range; very few protons penetrate beyond that distance.[4] Furthermore, the dose delivered to tissue is maximum just over the last few millimeters of the particle's range; this maximum is called the Bragg peak.[5]

To treat tumors at greater depths, the proton accelerator must produce a beam with higher energy, typically given in eV or electron volts. Proton therapy treats tumors closer to the surface of the body with lower energy protons. Accelerators used for proton therapy typically produce protons with energies in the range of 70 to 250 MeV (mega electron volts; million electron volts). Adjusting proton energy during the treatment maximizes the cell damage the proton beam causes within the tumor. Tissue closer to the surface of the body than the tumor receives reduced radiation, and therefore reduced damage. Tissues deeper in the body receive very few protons, so the dosage becomes immeasurably small.[4]

In most treatments, protons of different energies with Bragg peaks at different depths are applied to treat the entire tumor. These Bragg peaks are shown as thin blue lines in the figure to the right. The total radiation dosage of the protons is called the *spread-out Bragg peak* (SOBP), shown as a heavy dashed blue line in figure to the right. It is important to understand that, while tissues *behind* or *deeper than* the tumor receive no radiation from proton therapy, the tissues *in front of* or *shallower than* the tumor receive radiation dosage based on the SOBP.

11.2   History

The first suggestion that energetic protons could be an effective treatment method was made by Robert R. Wilson[6] in a paper published in 1946 while he was involved in the design of the Harvard Cyclotron Laboratory (HCL).[7] The first treatments were performed with particle accelerators built for physics research, notably Berkeley Radiation Laboratory in 1954 and at Uppsala in Sweden in 1957. In 1961, a collaboration began between HCL and the Massachusetts General Hospital (MGH) to pursue proton therapy. Over the next 41 years, this program refined and expanded these techniques while treating 9,116 patients[8] before the cyclotron was shut down in 2002. The world's first hospital-based proton therapy center was a low energy cyclotron centre for ocular tumours at the Clatterbridge Centre for Oncology in the UK, opened in 1989,[9] followed in 1990 at the Loma Linda University Medical Center (LLUMC) in Loma Linda, California. Later, The Northeast Proton Therapy Center at Massachusetts General Hospital was brought online, and the HCL treatment program was transferred to it during 2001 and 2002. By 2010 these facilities were joined by an additional seven regional hospital-based proton therapy centers in the United States alone, and many more worldwide.[10]

11.3   Application

Physicians use protons to treat conditions in two broad categories:

- Disease sites that respond well to higher doses of radiation, i.e., dose escalation. In some instances, dose escalation has demonstrated a higher probability of "cure" (i.e., local control) than conventional radiotherapy.[11] These include, among others, uveal melanoma (ocular tumors), skull base and paraspinal tumors (chondrosarcoma and chordoma), and unresectable sarcomas. In all these cases proton therapy achieves significant improvements in the probability of local control over conventional radiotherapy.[12][13][14] In treatment of ocular tumors, proton therapy also has high rates of maintaining the natural eye.[15]

The second broad class are those treatments where proton therapy's increased precision reduces unwanted side effects by lessening the dose to normal tissue. In these cases, the tumor dose is the same as in conventional therapy, so there is no expectation of an increased probability of curing the disease. Instead, the emphasis is on reducing the integral dose to normal tissue, thus reducing unwanted effects.[11]

Two prominent examples are pediatric neoplasms (such as medulloblastoma) and prostate cancer. In the case of pediatric treatments, a 2004 review gave theoretical advantages but did not report any clinical benefits.[16][17]

In prostate cancer cases, the issue is less clear. Some published studies found a reduction in long term rectal and genito-urinary damage when treating with protons rather than photons (meaning X-ray or gamma ray therapy). Others showed a small difference, limited to cases where the prostate is particularly close to certain anatomical structures.[18][19] The relatively small improvement found may be the result of inconsistent patient set-up and internal organ movement during treatment, which offsets most of the advantage of increased precision.[19][20][20][21] One source suggests that dose errors around 20% can result from motion errors of just 2.5 mm, and another that prostate motion is between 5–10 mm.[22]

However, the number of cases of prostate cancer diagnosed each year far exceeds those of the other diseases referred to above, and this has led some, but not all, facilities to devote a majority of their treatment slots to prostate treatments. For example, two hospital facilities devote roughly 65%[23] and 50%[24] of their proton treatment capacity to prostate cancer, while a third devotes only 7.1%.[25]

Overall worldwide numbers are hard to compile, but one example in the literature shows that in 2003 roughly 26% of proton therapy treatments worldwide were for prostate cancer.[26] Proton therapy for ocular (eye) tumors is a special case since this treatment requires only comparatively low energy protons (about 70 MeV). Owing to this low energy requirement, some particle therapy centers only treat ocular tumors.[8] Proton, or more generally, hadron therapy of tissue close to the eye affords sophisticated methods to assess the alignment of the eye that can vary significantly from other patient position verification approaches in image guided particle therapy.[27] Position verification and correction must ensure that the radiation spares sensitive tissue like the optic nerve to preserve the patient's vision.

11.4 Comparison with other treatments

The issue of when, whether, and how best to apply this technology is controversial.[28][29][30] As of 2012 there have been no controlled trials to demonstrate that proton therapy yields improved survival or other clinical outcomes (including impotence in prostate cancer) compared to other types of radiation therapy, although a five-year study of prostate cancer is underway at Massachusetts General Hospital.[31][32][33][34] Proton therapy is far more expensive than conventional therapy.[29][35] As of 2012 proton therapy required a very large capital investment (from US$100M to more than $180M).[28][30][36]

Preliminary results from a 2009 study, including high-dose treatments, showed very few side effects.[37]

NHS Choices has stated:

> We cannot say with any conviction that proton beam therapy is "better" overall than radiotherapy. (...) Some overseas clinics providing proton beam therapy heavily market their services to parents who are understandably desperate to get treatment for their children. Proton beam therapy can be very costly and it is not clear whether all children treated privately abroad are treated appropriately.[38][39]

11.4.1 X-ray radiotherapy

The figure at the right of the page shows how beams of X-rays (IMRT; left frame) and beams of protons (right frame), of different energies, penetrate human tissue. A tumor with a sizable thickness is covered by the IMRT spread out Bragg peak (SOBP) shown as the red lined distribution in the figure. The SOBP is an overlap of several pristine Bragg peaks (blue lines) at staggered depths.

Megavoltage X-ray therapy has less "skin scarring potential" than proton therapy: X-ray radiation at the skin, and at very small depths, is lower than for proton therapy. One study estimates that passively scattered proton fields have a slightly higher entrance dose at the skin (~75%) compared to therapeutic megavoltage (MeV) photon beams (~60%).[1] X-ray radiation dose falls off gradually, unnecessarily damaging tissue deeper in the body and damaging the skin and surface tissue opposite the beam entrance. The differences between the two methods depends on the:

- Width of the SOBP

- Depth of the tumor

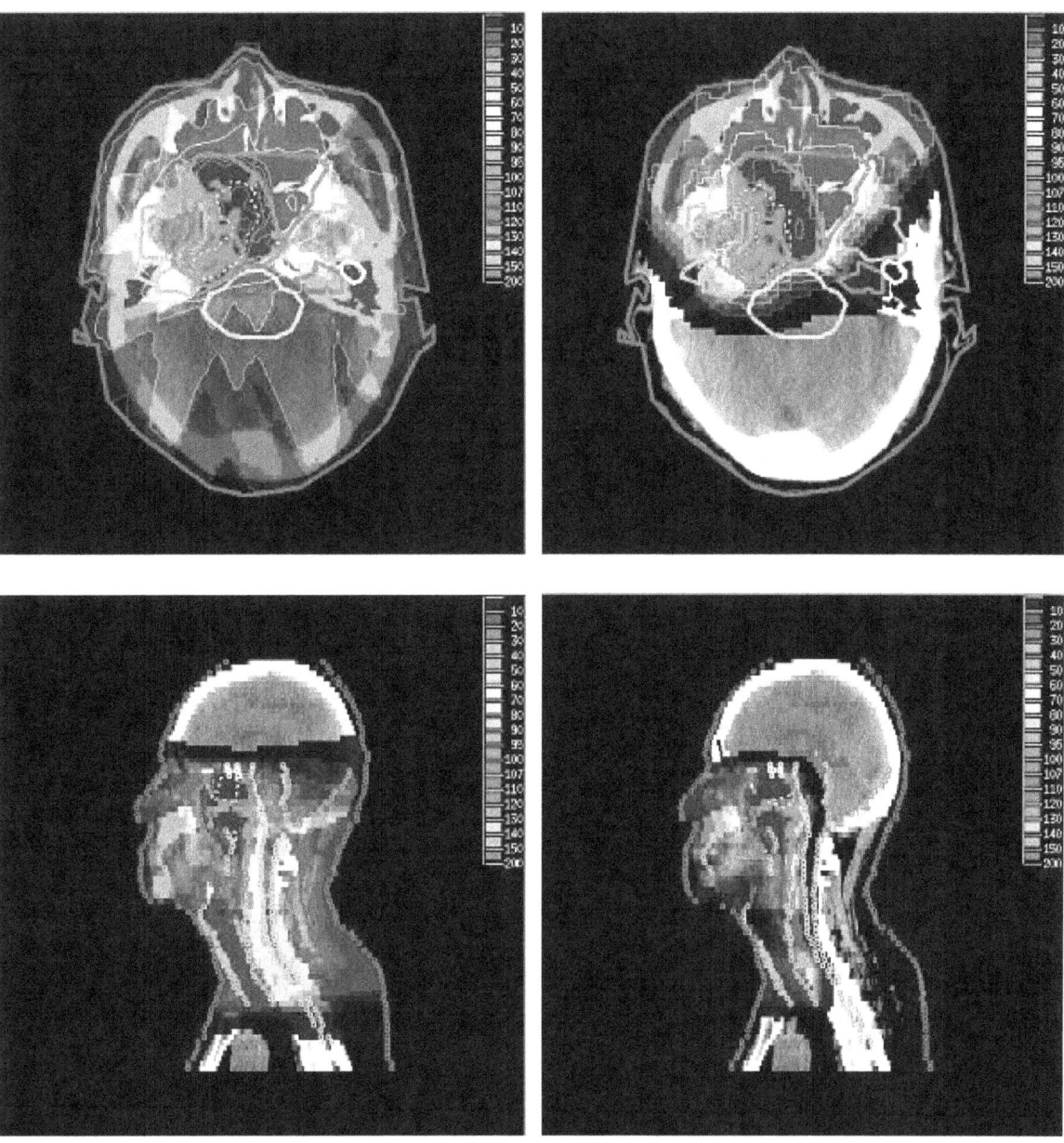

Irradiation of nasopharyngeal carcinoma by photon (X-ray) therapy (left) and proton therapy (right)

- Number of beams that treat the tumor

The X-ray advantage of reduced damage to skin at the entrance is partially counteracted by damage to skin at the exit point.

Since X-ray treatments are usually done with multiple exposures from opposite sides, each section of skin is exposed to both entering and exiting X-rays. In proton therapy, skin exposure at the entrance point is higher, but tissues on the opposite side of the body to the tumor receive no radiation. Thus, X-ray therapy causes slightly less damage to the skin and surface tissues, and proton therapy causes less damage to deeper tissues in front of and beyond the target.[3]

An important consideration in comparing these treatments is whether the equipment delivers protons via the scattering method (historically, the most common) or a spot scanning method. Spot scanning can adjust the width of the SOBP on a spot-by-spot basis, which reduces the volume of normal (healthy) tissue inside the high dose region. Also, spot scanning allows for intensity modulated proton therapy (IMPT), which determines individual spot intensities using an optimization algorithm that lets the user balance the competing goals of irradiating tumors while sparing normal tissue. Spot scanning availability depends on the machine and the institution. Spot scanning is more commonly known as pencil-beam scanning and is available on IBA, Hitachi, Mevion (known as hyperscan [40] and Not USFDA approved as of 2015) and Varian.

11.4.2 Surgery

Physicians base the decision to use surgery or proton therapy (or any radiation therapy) on the tumor type, stage, and location. In some instances, surgery is superior (e.g. cutaneous melanoma), in some instances radiation is superior (e.g., skull base chondrosarcoma), and in some instances they are comparable (e.g., prostate cancer). In some instances, they are used together (e.g., rectal cancer or early stage breast cancer). The benefit of external beam proton radiation lies in the dosimetric difference from external beam X-ray radiation and brachytherapy in cases where the use of radiation therapy is already indicated, rather than as a direct competition with surgery.[11] However, in the case of prostate cancer, the most common indication for proton beam therapy, no clinical study directly comparing proton therapy to surgery, brachytherapy, or other treatments has shown any clinical benefit for proton beam therapy. Indeed, the largest study to date showed that IMRT compared with proton therapy was associated with less gastrointestinal morbidity.[41]

11.5 Side effects and risks

Main articles: Radiation therapy § Side effects and Adverse effect

Proton therapy is a type of external beam radiotherapy, and shares risks and side effects of other forms of radiation therapy. However the dose outside of the treatment region can be significantly less for deep-tissue tumors than X-ray therapy, because proton therapy takes full advantage of the Bragg peak. Proton therapy has been in use for over 40 years, and is a mature treatment technology. However, as with all medical knowledge, understanding of the interaction of radiation (proton, X-ray, etc.) with tumor and normal tissue is still imperfect.[28]

11.6 Costs

Historically, proton therapy has been expensive. Goitein & Jermann's[42] analysis had previously determined the relative cost of proton therapy is approximately 2.4 times that of X-ray therapies. However, newer, more compact proton beam sources can be four to five times cheaper and offer more accurate three-dimensional targeting.[43][44] Thus the cost is expected to reduce as better proton technology becomes more widely available. A similar analysis by Lievens & Van den Bogaert[45] determined that the cost of proton therapy is not unrealistic and should not be the reason for denying patients access to this technology. In some clinical situations, proton beam therapy is clearly superior to the alternatives.[46][47] Another study in 2007 expressed concerns about the effectiveness of proton therapy for treating prostate cancer.[48] Although, with the advent of new developments in proton beam technology, such as improved scanning techniques and more precise dose delivery ('pencil beam scanning'), this situation may change considerably.[49] Amitabh Chandra, a health economist at Harvard University, has been quoted as saying that "Proton-beam therapy is like the death star of American medical technology... It's a metaphor for all the problems we have in American medicine."[50] However, another study has shown that proton therapy in fact brings cost savings.[51] The advent of second generation, and much less expensive, proton therapy equipment now being installed at various sites may change this picture significantly.[52]

11.7 Treatment centers

As of August 2013, 43 particle therapy facilities worldwide represented a total of 121 treatment rooms available to patients.[53] Of these, 28% are located in the US, 23% are located in Japan, and more than 96,537 patients had been treated.[54]

Control panel of the synchrocyclotron at the Orsay proton therapy center, France

One hindrance to universal use of the proton in cancer treatment is the size and cost of the cyclotron or synchrotron equipment necessary. Several industrial teams are working on development of comparatively small accelerator systems to deliver the proton therapy to patients.[55] Among the technologies being investigated are superconducting synchrocyclotrons (also known as FM Cyclotrons), ultra-compact synchrotrons, dielectric wall accelerators,[55] and linear particle accelerators.[44]

11.7.1   United States

Proton treatment centers in the United States as of 2014 (in chronological order of first treatment date) include:[9][56]

The Mayo Clinic is soon launching their own proton therapy centers, with the Minnesota campus expecting to begin treating patients in late June 2015, and the Arizona campus opening its doors sometime in 2016.[70]

11.7.2   Outside the USA

11.8   United Kingdom

In 2013 the British government announced that £250 million had been budgeted to establish two centers for advanced radiotherapy, to open in 2018 at the Christie Hospital NHS Foundation Trust in Manchester and University College London Hospitals NHS Foundation Trust. These would offer high-energy proton therapy, currently unavailable in the UK, as well as other types of advanced radiotherapy including intensity modulated radiotherapy (IMRT) and image guided radiotherapy (IGRT).[73] In 2014, only low-energy proton therapy was available in the UK, at the Clatterbridge Cancer Centre NHS Foundation Trust in Merseyside. But NHS England has paid to have suitable cases treated abroad, mostly in the US. Such cases have risen from 18 in 2008 to 122 in 2013, 99 of whom were children. The cost to the National Health Service averaged around £100,000 per case.[74]

In January 2015, it was announced the UK would get its first high energy proton beam therapy centre a year earlier than expected.[75] Private sector firm, Advanced Oncotherapy, signed a deal with Howard de Walden Estates Ltd to install a machine in London's Harley Street to come on stream in 2017. Advanced Oncotherapy's system allows for more rapid movement and energy variation of the proton beam than is currently available from existing technologies.[76] The technological advance also allows the company to build proton therapy facilities one-third smaller and one-fifth the cost of facilities based on the NHS first-generation machines which will come on stream a year later. The NHS has been criticised by leading doctors for buying old equipment.[77]

11.9 See also

- Particle therapy, Charged particle therapy, Hadron, Microbeam

- Fast neutron therapy

- Boron neutron capture therapy

- Linear energy transfer, Electromagnetic radiation and health

- Dosimetry, Dosimeter, Ionizing radiation

- List of oncology-related terms

11.10 References

[1] Adapted from "Proton beam therapy" W P Levin, H Kooy, J S Loeffler and T F DeLaney British Journal of Cancer (2005) 93, 849–854

[2] O. Jakel: State of the art in hadron therapy. AIP Conference Proceedings, vol. 958, no.1, 2007, pp. 70-77

[3] "Zap! You're not dead." *Economist*, 8 September 2007. **384** (8545):13-14

[4] Metz, James (2006-07-31). "Differences Between Protons and X-rays". The Abramson Cancer Center of the University of Pennsylvania. Retrieved 2008-02-04. the beam then stops, resulting in virtually no radiation to the tissue beyond the target- or no "exit dose"

[5] Camphausen KA, Lawrence RC. "Principles of Radiation Therapy" in Pazdur R, Wagman LD, Camphausen KA, Hoskins WJ (Eds) Cancer Management: A Multidisciplinary Approach. 11 ed. 2008.

[6] "Radiological Use of Fast Protons", R. R. Wilson, Radiology, 47:487-491 (1946)

[7] Richard Wilson, "A Brief History of the Harvard University Cyclotrons", Harvard University Press, 2004, pp 9

[8] "PTCOG: Particle Therapy Co-Operative Group". Ptcog.web.psi.ch. Retrieved 2009-09-03.

[9] "Particle therapy facilities in operation". Particle Therapy Co-Operative Group. 2013-08-27. Retrieved 2014-09-01.

[10] "Particle therapy facilities in operation". Particle Therapy Co-Operative Group. Retrieved 2010-04-27.

[11] R. P. Levy et al., The current status and future directions of heavy charged particle therapy in medicine, *AIP Journal*, March 2009

[12] Hug E. B. et al. (1999). ": Proton radiation therapy for chordomas and chondrosarcomas of the skull base". *J. Neurosurgery* **91**: 432–439. doi:10.3171/jns.1999.91.3.0432.

[13] E. Gragoudas et al., Evidence-based estimates of outcomes in patients treated for intraocular melenoma", *Arch. Ophthalmol.*120, 1665-1671 (2002)

[14] Munzenrider J. E., Liebsch N. J. (1999). "Proton radiotherapy for tumors of the skull base". *Strahnlenther. Onkol* **175**: 57–63. doi:10.1007/bf03038890.

[15] "Proton Therapy for Ocular Tumors".

[16] W. H. St. Clair et al, Advantage of protons compared to conventional X-ray or IMRT in the treatment of a pediatric patient with medulloblastoma, *Int. J. Radiat. Oncol. Biol. Phys.*58, 727-734 (2004)

[17] D.G. Kirsch and N. J. Tarbell, Conformal radiation therapy for childhood CNS tumors, *The Oncologist* 9(4), 442-450 (2004)

[18] Slater J. D. et al. (2004). ", Proton therapy for prostate cancer; the initial Loma Linda University experience". *Int. J. Radiat. Oncol. Biol. Phys* **59**: 348–352.

[19] A. L. Zietman et al, Comparisons of conventional-dose vs. high-dose conformal radiation therapy in clinically localized adeno-carcinoma of the prostate: a randomized controlled trial, *J. A. M. A.* 294(10) 1233-1239 (2005)

[20] R. deCrevoisier et al, "Increased risk of biochemical and local failure in patients with distended rectum on the planning CT for prostate cancer radiotherapy," *Int. J. Radiat. Oncol. Biol. Phys.* 62(4) 965-973 (2005)

[21]Lambert et al. (2005). "Intrafractional motion during proton beam scanning".*Phys.Med.Biol.***50**:4853–4862.doi:10.1088/00-9155/50/20/008.

[22] Byrne Thomas E (2005). "A Review of Prostate Motion with Considerations for the Treatment of Prostate Cancer". *Medical Dosimerty* **30** (3): 155–161. doi:10.1016/j.meddos.2005.03.005.

[23] Dyk, Jacob, Van (1999). *The modern technology of radiation oncology: A Compendium for Medical Physicists and Radiation Oncologists.* p. 826: Medical Physics Publishing Corporation. p. 1072. ISBN 9780944838389. Proton Patient Summary - Inception Through December 1998...Prostate...2591 64.3%

[24] "The Promise of Proton-Beam Therapy". U.S. News and World Report. 2008-04-16. Retrieved 2008-02-20.

[25] "Francis H. Burr Proton Therapy Center" (PDF).

[26] J. Sisterson, Ion beam therapy in 2004, *Nuclear Instruments and Methods in Physics Research* B 241 713-716 (2005)

[27] Boris Peter Selby et al., "Pose estimation of eyes for particle beam treatment of tumors.", Bildverarbeitung für die Medizin (Medical Image Processing); Munich, 2007 pp. 368-373.

[28] Joel E. Tepper, MD, and A. William Blackstock, MD (20 October 2009). "EDITORIAL: Randomized Trials and Technology Assessment". *Annals of Internal Medicine* **151** (8): 583–584. doi:10.7326/0003-4819-151-8-200910200-00146. PMID 19755346.

[29] BOULTON, GUY (March 30, 2008). "High cost; of high tech; Outlay vs. benefit of expensive medical devices questioned". Milwaukee Journal Sentinel. Retrieved 2009-09-03. Despite that controversy, roughly a dozen proton therapy centers have been proposed throughout the country, including northern Illinois

[30] David Whelan and Robert Langreth (March 16, 2009). "The $150 Million Zapper:Does every cancer patient really need proton-beam therapy?". *Forbes*. Retrieved 2009-09-03.

[31] Terasawa, T; Teruhiko Terasawa, MD; Tomas Dvorak, MD; Stanley Ip, MD; Gowri Raman, MD; Joseph Lau, MD; Thomas A. Trikalinos, MD, PhD (20 October 2009). "Systematic Review: Charged-Particle Radiation Therapy for Cancer". *Annals of Internal Medicine* **151** (8): 556–565. doi:10.7326/0003-4819-151-8-200910200-00145. PMID 19755348.[FREE]

[32] Proton beams vs. radiation: 5-year MGH study seeks definitive answers about costly prostate cancer treatment, By Carolyn Y. Johnson, Boston Globe, May 14, 2012

[33] "Particle Beam Radiation Therapies for Cancer: Policymaker Summary Guide". U.S. Department of Health and Human Services. September 14, 2009. Retrieved 2009-10-09.

[34] "Particle Beam Radiation Therapies for Cancer Final Research Review". U.S. Department of Health and Human Services Federal Agency for Healthcare Research and Quality. September 14, 2009.

[35] Feldstein, Dan (Oct 23, 2005). "M.D. Anderson private venture raises questions/ Proton-therapy benefits at center won't merit costs of care, some say". Houston Chronicle. Retrieved 2009-10-01. M.D. Anderson officials estimate that when patients on all types of insurance and payment plans are mixed together, proton delivery will cost an average of $37,000 per patient for prostate treatment, compared with $29,000 for IMRT and $21,000 for standard radiation. The amount excludes doctor fees, which will be roughly the same for each.

[36] Emanuel EJ; Pearson SD (2012-01-02). "It Costs More, but Is It Worth More?". *The Opinion Pages.* New York: The New York Times. Retrieved 2012-01-03.

[37] Cox, Jeremy (2009-11-23). "UF Proton Therapy Institute study shows positive outcomes". Jacksonville.com. Retrieved 2009-12-22.

[38] Ashya King: This story isn't quite what it seems

[39] What is proton beam therapy?

[40] "Introducing Hyperscan". Mevion Medical Systems. 2015-04-19.

[41] Sheets, NC; Goldin, GH; Meyer, AM; Wu, Y; Chang, Y; Stürmer, T; Holmes, JA; Reeve, BB; Godley, PA; Carpenter, WR; Chen, RC (Apr 18, 2012). "Intensity-modulated radiation therapy, proton therapy, or conformal radiation therapy and morbidity and disease control in localized prostate cancer.". *JAMA: the Journal of the American Medical Association* **307** (15): 1611–20. doi:10.1001/jama.2012.460. PMID 22511689.

[42] Goitein, M., & Jermann, M. 2003. "The Relative Costs of Proton and X-ray Radiation Thearpy. (2003)". *Clinical Oncology, 15, S37–50* **15**: S37–S50. doi:10.1053/clon.2002.0174.

[43] "Siteman Cancer Center Treats First Patient With First-of-Its-Kind Proton Therapy System". PRWeb. Retrieved 2014-01-09.

[44] "God particle technology to cancer patients".

[45] Lievens, Y., & Van den Bogaert, W. 2005. Proton beam therapy: Too expensive to become true? Radiotherapy and Oncology, 75, 131–3.

[46] St Clair, W. H., Adams, J. A., Bues, M., Fullerton, B. C., La Shell, S., Kooy, H. M., Loeffler, J. S., and Tarbell, N. J. (2004). Advantage of protons compared to conventional X-ray or IMRT in the treatment of a pediatric patient with medulloblastoma. Int. J. Radiat. Oncol. Biol. Phys. 58, 727–734.

[47] Merchant, T. E., Hua, C. H., Shukla, H., Ying, X., Nill, S., and Oelfke, U. (2008). Proton versus photon radiotherapy for common pediatric brain tumors: comparison of models of dose characteristics and their relationship to cognitive function. Pediatr. Blood Cancer 51, 110–117.

[48] Konski A., Speier W., Hanlon A., Beck J. R., Pollack A. (2007). "Is proton beam therapy cost effective in the treatment of adenocarcinoma of the prostate?". *J. Clin. Oncol* **25**: 3603–3608. doi:10.1200/jco.2006.09.0811.

[49] NGUYEN P., TROFIMOV A. et al. (2008). "Proton-Beam vs. Intensity-Modulated Radiation Therapy, Which Is Best for Treating Prostate Cancer?". *Oncology* **22**: 7.

[50] Langreth, Robert (March 26, 2012). "Prostate Cancer Therapy Too Good to Be True Explodes Health Cost". Bloomberg. Retrieved May 16, 2013.

[51] Lundkvist J., Ekman M., Ericsson S. R., Jönsson B., Glimelius B. (2005a). "Cost-effectiveness of proton radiation in the treatment of childhood medulloblastoma". *Cancer* **103**: 793–801. doi:10.1002/cncr.20844.

[52] "Mevion Medical Systems continues Manufacturing Ramp Up". Marketwatch.com. 2012-11-01. Retrieved 2012-11-29.

[53] CSIntell (2013-09-11). "Proton Therapy World Market to nearly triple by 2018". PRLog. Retrieved 2013-11-05.

[54] "Hadron Therapy Patient Statistics" (PDF). Particle Therapy Co-Operative Group. Retrieved 2012-03-30.

[55] J.N.A. Matthews: "Accelerators shrink to meet growing demand for proton therapy", Physics Today, March 2009, p. 22

[56] "N.J. proton therapy center opens today". DotMed.com. Retrieved 2012-03-30.

[57] http://cyclotron.crocker.ucdavis.edu/

[58] "Proton Therapy Treatment and Research Center". Loma Linda University Medical Center. Retrieved 2013-11-05.

[59] "Proton Therapy Jacksonville | Cancer Treatment". University of Florida Proton Therapy Institute. Retrieved 2013-11-05.

[60] "Proton Therapy Center". University of Texas MD Anderson Cancer Center. Retrieved 2013-11-05.

[61] "Oklahoma Proton Therapy Treatment Center". ProCure. Retrieved 2013-11-05.

[62] "Illinois Proton Therapy Treatment Center". ProCure. Retrieved 2013-11-05.

[63] "Proton Therapy at Penn Medicine". Perelman Center for Advanced Medicine. Retrieved 2013-11-05.

[64] "New Jersey Proton Therapy Treatment Center". ProCure. Retrieved 2013-11-05.

[65] "Elegant and Precise". Mevion Medical Systems. Retrieved 2015-04-19.

[66] "Introducing the Mevion S250". Mevion. Retrieved 2015-04-19.

[67] "Proton therapy cancer treatment center opens, first of its kind in Tennessee". WATE TV. Retrieved 2014-01-25.

[68] "Scripps Proton Therapy Center". Scripps Health. Retrieved 2015-03-28.

[69] "Oncology, Solutions, Proton Therapy". Varian Medical Systems. Retrieved 2015-04-19.

[70] "Questions and Answers". *Proton Beam Therapy*. Mayo Clinic. Retrieved 21 April 2015.

[71] TRIUMF Proton Therapy

[72] Clatterbridge Cancer Centre NHS Foundation Trust

[73] "Manchester and London proton beam therapy units confirmed", Press release, Press Association, Cancer Research UK, 1 August 2013

[74] "Ashya King case: What is proton beam therapy?" BBC news story with NHS England figures, 31 August 2014

[75] http://www.lse.co.uk/AllNews.asp?code=dqaddbqb&headline=Advanced_Oncotherapy_Signs_Lease_Deal_To_Establish_LIGHT_Centre

[76] "NeoStem (Amex: NBS) 15M units Prices at $0.40 per unit for $6M Public Offering". www.proactiveinvestors.co.uk. Retrieved 11 August 2015.

[77] "Brain tumor boy Ashya King 'could soon be walking unaided'... as NHS is blasted for buying 'old' equipment". The Mail on Sunday. Retrieved 11 August 2015.

11.11 Further reading

- Greco C., Wolden S. (Apr 2007). "Current status of radiotherapy with proton and light ion beams". *Cancer* **109** (7): 1227–38. doi:10.1002/cncr.22542. PMID 17326046.

- "Use of Protons for Radiotherapy", A.M. Koehler, Proc. of the Symposium on Pion and Proton Radiotherapy, Nat. Accelerator Lab., (1971).

- A.M. Koehler, W.M. Preston, "Protons in Radiation Therapy: comparative Dose Distributions for Protons, Photons and Electrons *Radiology* 104(1):191–195 (1972).

- "Bragg Peak Proton Radiosurgery for Arteriovenous Malformation of the Brain" R.N. Kjelberg, presented at First Int. Seminar on the Use of Proton Beams in Radiation Therapy, Moskow (1977).

- Austin-Seymor, M.J. Munzenrider, et al. "Fractionated Proton Radiation Therapy of Cranial and Intracrainial Tumors" *Am. J. of Clinical Oncology* 13(4):327–330 (1990).

- "Proton Radiotherapy", Hartford, Zietman, et al. in *Radiotheraputic Management of Carcinoma of the Prostate*, A. D'Amico and G.E. Hanks. London,UK, Arnold Publishers: 61–72 (1999).

11.12 External links

- Proton therapy—MedlinePlus Medical Encyclopedia

- Proton Therapy

- "Proton therapy is coming to the UK, but what does it mean for patients?", Arney, Kat, Science blog, Cancer Research UK, 16 September 2013

- CERN - AVO

- Proton Therapy in Korea

Chapter 12

Hydron (chemistry)

In chemistry, a **hydron** is the general name for a cationic form of atomic hydrogen, represented with the symbol H+
. The term **"proton"** refers to the cation of protium, the most common isotope of hydrogen. The term "hydron" includes cations of hydrogen regardless of their isotopic composition: thus it refers collectively to protons ($^1H^+$) for the protium isotope, deuterons ($^2H^+$ or D^+) for the deuterium isotope, and tritons ($^3H^+$ or T^+) for the tritium isotope. Unlike other ions, the hydron consists only of a bare atomic nucleus.

The hydron (a completely free or "naked" hydrogen atomic nucleus) is too reactive to occur in many liquids, even though it is sometimes visualized to do so by students of chemistry. A free hydron would react with a molecule of the liquid to form a more complicated cation. Examples are the hydronium ion in water-based acids, and H
2F+
, the unstable cation of fluoroantimonic acid, the strongest superacid. For this reason, in such liquids including liquid acids, hydrons diffuse by contact from one complex cation to another, via the Grotthuss mechanism.[2]

The hydrated form of the hydrogen cation, the hydronium (hydroxonium) ion H
3O+
(aq), is a key object of Arrhenius' definition of acid. Other hydrated forms, the Zundel cation H
5O+
2 which is formed from a proton and two water molecules, and the Eigen cation H
9O+
4, a hydronium ion and three water molecules, play an important role in "hydron hopping" according to the Grotthuss mechanism. The hydron itself is crucial in more general Brønsted–Lowry acid–base theory, which extends the concept of acid–base chemistry beyond aqueous solutions.

The negatively charged counterpart of the hydron is the hydride anion, H–
.

12.1   Isotopes of hydron

1. Proton, having the symbol p or $^1H^+$, is the +1 ion of protium, ^1H.

2. Deuteron, having the symbol $^2H^+$ or D^+, is the +1 ion of deuterium, ^2H or D.

3. Triton, having the symbol $^3H^+$ or T^+, is the +1 ion of tritium, ^3H or T.

Other isotopes of hydrogen are too unstable to be relevant in chemistry.

12.2 History of the term

The term "hydron" is recommended by IUPAC to be used instead of "proton" if no distinction is made between the isotopes proton, deuteron and triton, all found in naturally occurring undifferentiated isotope mixtures. The name "proton" refers to isotopically pure $^1H^+$.[3] On the other hand, referring to the hydron as simply *hydrogen ion* is not recommended because hydrogen anions also exist.[4]

The term "hydron" was defined by IUPAC in 1988.[5][6] Traditionally, the term "proton" was and is used in place of "hydron". The latter term is generally only used in the context where comparisons between the various isotopes of hydrogen is important (as in the kinetic isotope effect or hydrogen isotopic labeling). Otherwise, referring to hydrons as protons is still considered acceptable, for example in such terms as protonation, deprotonation, proton pump or proton channel. The transfer of H+
in an acid-base reaction is usually referred to as *proton* transfer. Acid and bases are referred to as *proton* donors and acceptors correspondingly.

However, although 99.9844% of natural hydrogen nuclei are protons, the remainder (about 156 per million in sea water) are deuterons (see deuterium). A rare triton also occurs (see tritium).

12.3 See also

- Deprotonation

- Superacid

- Dihydrogen cation

- Trihydrogen cation

- Hydrogen ion cluster

12.4 References

[1] "hydron (CHEBI:15378)". *Chemical Entities of Biological Interest (ChEBI)*. UK: European Bioinformatics Institute.

[2] Computer modeling of proton-hopping in superacids.

[3] Nomenclature of Inorganic Chemistry-IUPAC Recommendations 2005 Red Book 2005.pdf IR-3.3.2, p.48

[4] *Compendium of Chemical Terminology*, 2nd edition McNaught, A.D. and Wilkinson, A. Blackwell Science, 1997 [ISBN 0-86542-684-8], also online

[5] IUPAC, *Compendium of Chemical Terminology*, 2nd ed. (the "Gold Book") (1997). Online corrected version: (2006–) "hydron".

[6] Bunnet, J.F.; Jones, R.A.Y. (1968). "Names for hydrogen atoms, ions, and groups, and for reactions involving them (Recommendations 1988)" (PDF). *Pure Appl. Chem.* **60** (7): 1115–6. doi:10.1351/pac198860071115.

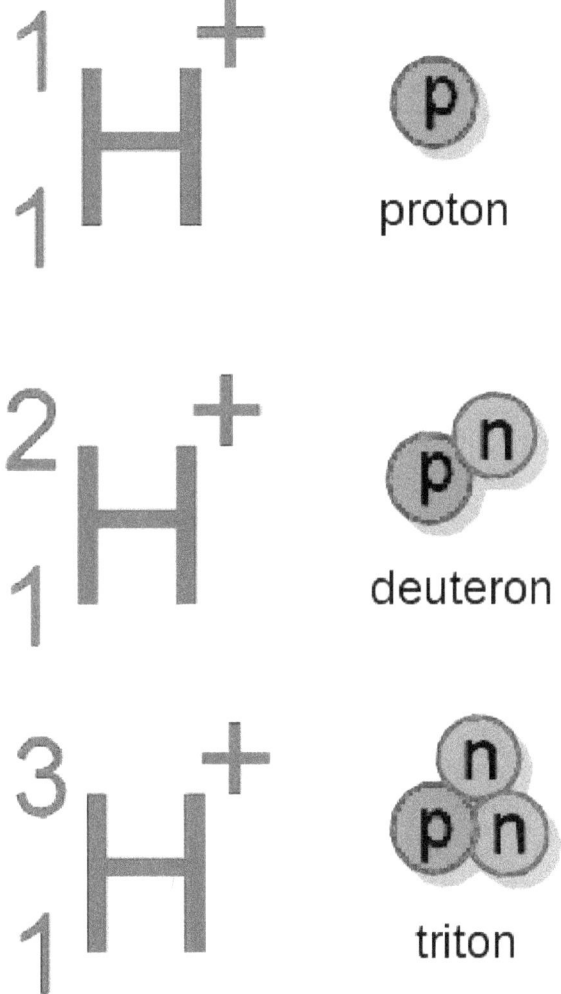

The three isotopes of the hydrogen cation (hydron).

Chapter 13

Proton nuclear magnetic resonance

1D PROTON SPECTRUM

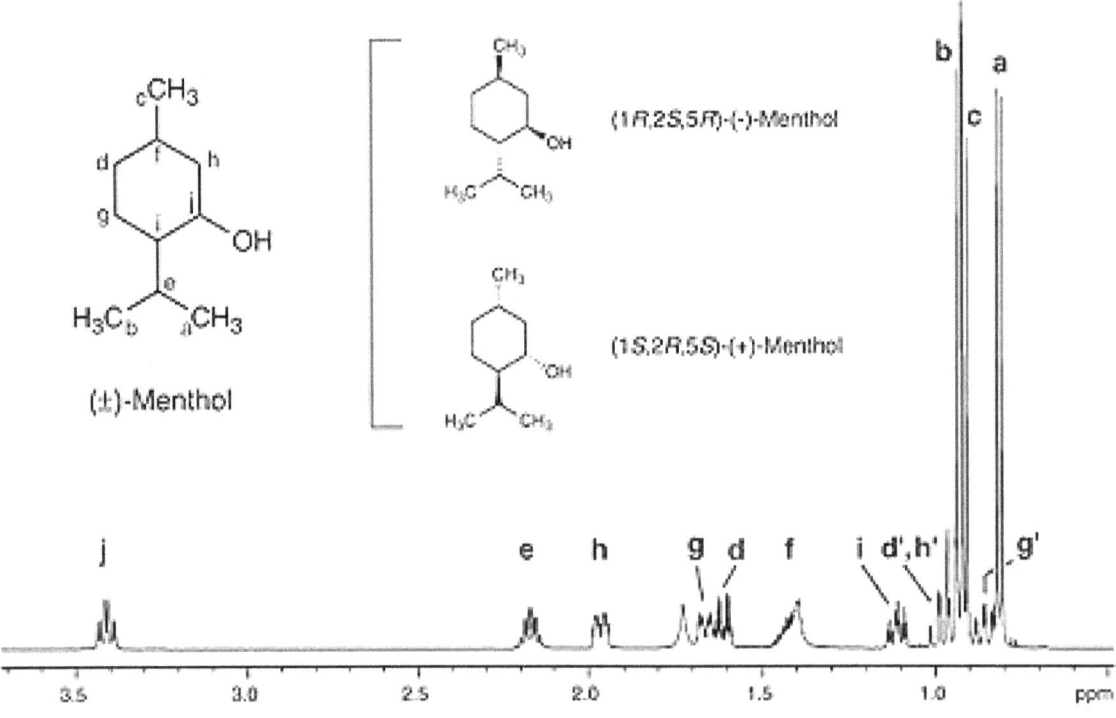

Example [^1]*H NMR spectrum (1-dimensional) of a mixture of menthol enantiomers plotted as signal intensity (vertical axis) vs. chemical shift (in ppm on the horizontal axis). Signals from spectrum have been assigned hydrogen atom groups (a through j) from the structure shown at upper left.*

Proton nuclear magnetic resonance (**proton NMR, hydrogen-1 NMR**, or [^1]**H NMR**) is the application of nuclear magnetic resonance in NMR spectroscopy with respect to hydrogen-1 nuclei within the molecules of a substance, in order to determine the structure of its molecules.[1] In samples where natural hydrogen (H) is used, practically all the hydrogen consists of the isotope [^1]H (hydrogen-1; i.e. having a proton for a nucleus). A full [^1]H atom is called protium.

Simple NMR spectra are recorded in solution, and solvent protons must not be allowed to interfere. Deuterated (deuterium = [^2]H, often symbolized as D) solvents especially for use in NMR are preferred, e.g. deuterated water, D_2O, deuterated acetone, $(CD_3)_2CO$, deuterated methanol, CD_3OD, deuterated dimethyl sulfoxide, $(CD_3)_2SO$, and deuterated chloroform, $CDCl_3$. However, a solvent without hydrogen, such as carbon tetrachloride, CCl_4 or carbon disulphide, CS_2, may also be used.

Historically, deuterated solvents were supplied with a small amount (typically 0.1%) of tetramethylsilane (TMS) as an internal standard for calibrating the chemical shifts of each analyte proton. TMS is a tetrahedral molecule, with all protons being chemically equivalent, giving one single signal, used to define a chemical shift = 0 ppm. [2] It is volatile, making sample recovery easy as well. Modern spectrometers are able to reference spectra based on the residual proton in the solvent (e.g. the $CHCl_3$, 0.01% in 99.99% $CDCl_3$). Deuterated solvents are now commonly supplied without TMS.

Deuterated solvents permit the use of deuterium frequency-field lock (also known as deuterium lock or field lock) to offset the effect of the natural drift of the NMR's magnetic field B_0 . In order to provide deuterium lock, the NMR constantly monitors the deuterium signal resonance frequency from the solvent and makes changes to the B_0 to keep the resonance frequency constant.[3] Additionally, the deuterium signal may be used to accurately define 0 ppm as the resonant frequency of the lock solvent and the difference between the lock solvent and 0 ppm (TMS) are well known.

Proton NMR spectra of most organic compounds are characterized by chemical shifts in the range +14 to −4 ppm and by spin-spin coupling between protons. The integration curve for each proton reflects the abundance of the individual protons.

Simple molecules have simple spectra. The spectrum of ethyl chloride consists of a triplet at 1.5 ppm and a quartet at 3.5 ppm in a 3:2 ratio. The spectrum of benzene consists of a single peak at 7.2 ppm due to the diamagnetic ring current.

Together with Carbon-13 NMR, proton NMR is a powerful tool for molecular structure characterization.

13.1   Chemical shifts

Chemical shift values, symbolized by δ, are not precise, but typical - they are to be therefore regarded mainly as a reference. Deviations are in ±0.2 ppm range, sometimes more. The exact value of chemical shift depends on molecular structure and the solvent, temperature, magnetic field in which the spectrum is being recorded and other neighboring functional groups. Hydrogen nuclei are sensitive to the hybridization of the atom to which the hydrogen atom is attached and to electronic effects. Nuclei tend to be deshielded by groups which withdraw electron density. Deshielded nuclei resonate at higher δ values, whereas shielded nuclei resonate at lower δ values.

Examples of electron withdrawing substituents are -OH, -OCOR, -OR, $-NO_2$ and halogens. These cause a downfield shift of approximately 2-4 ppm for H atoms on $C\alpha$ and of less than 1-2 ppm for H atoms on C_β. $C\alpha$ is an aliphatic C atom directly bonded to the substituent in question, and C_β is an aliphatic C atom bonded to $C\alpha$. Carbonyl groups, olefinic fragments and aromatic rings contribute sp^2 hybridized carbon atoms to an aliphatic chain. This causes a downfield shift of 1-2 ppm at $C\alpha$.

Note that labile protons (-OH, $-NH_2$, -SH) have no characteristic chemical shift. However such resonances can be identified by the disappearance of a peak when reacted with D_2O, as deuterium will replace a protium atom. This method is called a **D_2O shake**. Acidic protons may also be suppressed when a solvent containing acidic deuterium ions (e.g. methanol-d_4) is used. An alternate method for identifying protons that are not attached to carbons is the heteronuclear single quantum coherence (HSQC) experiment, which correlates protons and carbons that are one bond away from each other. A hydrogen that is not attached to a carbon can be identified because it does not have a crosspeak in the HSQC spectrum.

13.2   Signal strength

Signal strength in proton NMR is the intensity of signals displayed in a NMR spectrum and is proportional to the molar concentration of the corresponding proton.[4] The intensity is usually expressed as the integration of area under a particular NMR signal peak.

Signal strength is a relative value that depends on both the chemical compounds analyzed as well as the instrument and conditions used. Usually the signal strengths of different protons in one NMR spectrum are used to calculate their ratio. Comparison of the signal strength of protons in different NMR spectrum is usually not useful.

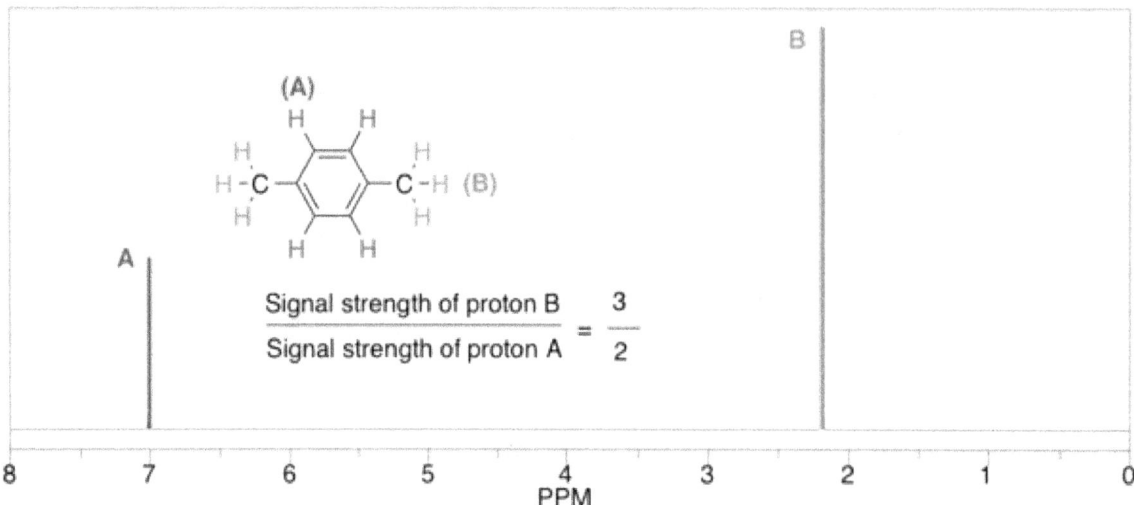

Predicted proton NMR of 1,4-dimethylbenzene from ChemDraw. The ratio of signal strengths of proton A and proton B equals to their molar ratio in the molecule.

Signal strength and chemical shift are two parameters that are used in combination for analysis of samples by proton NMR. The signal strength of protons from different molecules can be used to determine their molar ratio in a sample mixture, while the signal strength of different sets of protons from the same molecule determines their relative ratio. For example, p-Xylene (1,4-dimethylbenzene) has two sets of protons (proton A and proton B) that have different chemical environment. The proton NMR spectrum of p-Xylene shows two sets of protons with signal strength ratio of 3:2 that is same as the molar ratio of those two sets of protons in the molecule.

13.3 Spin-spin couplings

The chemical shift is not the only indicator used to assign a molecule. Because nuclei themselves possess a small magnetic field, they influence each other, changing the energy and hence frequency of nearby nuclei as they resonate—this is known as spin-spin coupling. The most important type in basic NMR is *scalar coupling*. This interaction between two nuclei occurs through chemical bonds, and can typically be seen up to three bonds away.

The effect of scalar coupling can be understood by examination of a proton which has a signal at 1ppm. This proton is in a hypothetical molecule where three bonds away exists another proton (in a CH-CH group for instance), the neighbouring group (a magnetic field) causes the signal at 1 ppm to split into two, with one peak being a few hertz higher than 1 ppm and the other peak being the same number of hertz lower than 1 ppm. These peaks each have half the area of the former **singlet** peak. The magnitude of this splitting (difference in frequency between peaks) is known as the coupling constant. A typical coupling constant value would be 7 Hz.

The coupling constant is independent of magnetic field strength because it is caused by the magnetic field of another nucleus, not the spectrometer magnet. Therefore it is quoted in hertz (frequency) and not ppm (chemical shift).

In another molecule a proton resonates at 2.5 ppm and that proton would also be split into two by the proton at 1 ppm. Because the magnitude of interaction is the same the splitting would have the same coupling constant 7 Hz apart. The spectrum would have two signals, each being a **doublet**. Each doublet will have the same area because both doublets are produced by one proton each.

The two doublets at 1 ppm and 2.5 ppm from the fictional molecule CH-CH are now changed into CH_2-CH:

- The total area of the 1 ppm CH_2 peak will be twice that of the 2.5 ppm CH peak.

- The CH_2 peak will be split into a doublet by the CH peak—with one peak at 1 ppm + 3.5 Hz and one at 1 ppm -

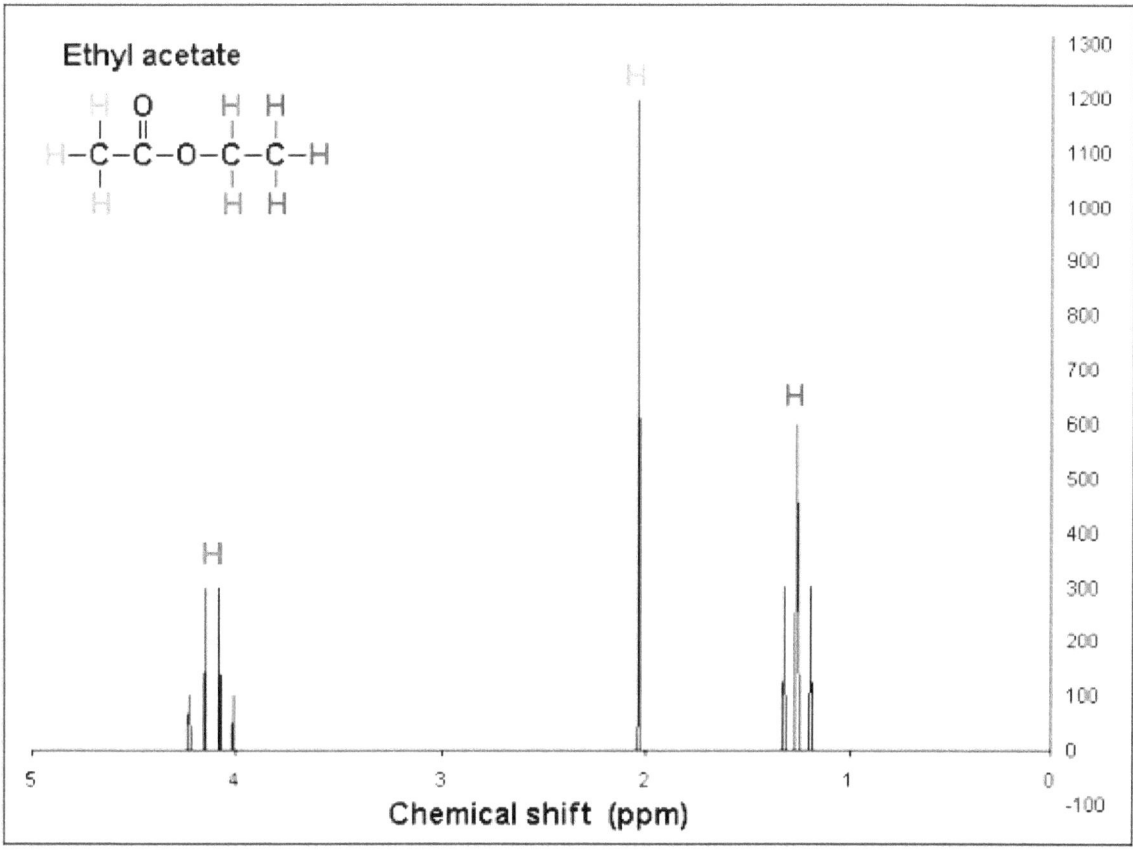

Example ^1H NMR spectrum *(1-dimensional) of ethyl acetate plotted as signal intensity vs. chemical shift. There are three different types of H atoms in ethyl acetate regarding NMR. The hydrogens (H) on the CH_3COO- (acetate) group are not coupling with the other H atoms and appear as a singlet, but the $-CH_2$- and $-CH_3$ hydrogens of the ethyl group ($-CH_2CH_3$) are coupling with each other, resulting in a quartet and triplet respectively.*

     3.5 Hz (total splitting or coupling constant is 7 Hz).

In consequence the CH peak at 2.5 ppm will be split *twice* by each proton from the CH_2. The first proton will split the peak into two equal intensities and will go from one peak at 2.5 ppm to two peaks, one at 2.5 ppm + 3.5 Hz and the other at 2.5 ppm - 3.5 Hz—each having equal intensities. However these will be split again by the second proton. The frequencies will change accordingly:

- The 2.5 ppm **+** 3.5 Hz signal will be split into 2.5 ppm + 7 Hz and 2.5 ppm

- The 2.5 ppm **-** 3.5 Hz signal will be split into 2.5 ppm and 2.5 ppm - 7 Hz

The net result is not a signal consisting of 4 peaks but three: one signal at 7 Hz above 2.5 ppm, two signals occur at 2.5 ppm, and a final one at 7 Hz below 2.5 ppm. The ratio of height between them is 1:2:1. This is known as a **triplet** and is an indicator that the proton is three-bonds from a CH_2 group.

This can be extended to any CH_n group. When the CH_2-CH group is changed to CH_3-CH_2, keeping the chemical shift and coupling constants identical, the following changes are observed:

- The relative areas between the CH_3 and CH_2 subunits will be 3:2.

- The CH_3 is coupled to two protons into a 1:2:1 **triplet** around 1 ppm.

- The CH_2 is coupled to *three* protons.

Something split by three identical protons takes a shape known as a **quartet**, each peak having relative intensities of 1:3:3:1.

A peak is split by n identical protons into components whose sizes are in the ratio of the nth row of Pascal's triangle:

n 0 singlet 1 1 doublet 1 1 2 triplet 1 2 1 3 quartet 1 3 3 1 4 quintet 1 4 6 4 1 5 sextet 1 5 10 10 5 1 6 septet 1 6 15 20 15 6 1 7 octet 1 7 21 35 35 21 7 1 8 nonet 1 8 28 56 70 56 28 8 1

Because the nth row has $n+1$ components, this type of splitting is said to follow the "$n+1$ rule": a proton with n neighbors appears as a cluster of $n+1$ peaks.

With 2-methylpropane, $(CH_3)_3CH$, as another example: the CH proton is attached to three identical methyl groups containing a total of 9 identical protons. The C-H signal in the spectrum would be split into **ten** peaks according to the (n + 1) rule of multiplicity. Below are NMR signals corresponding to several simple multiplets of this type. Note that the outer lines of the nonet (which are only 1/8 as high as those of the second peak) can barely be seen, giving a superficial resemblance to a septet.

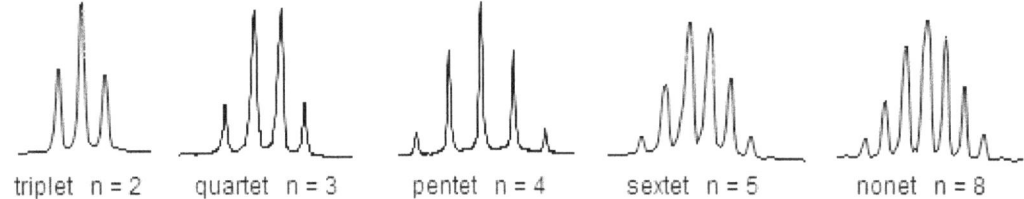

triplet n = 2   quartet n = 3   pentet n = 4   sextet n = 5   nonet n = 8

When a proton is coupled to two different protons, then the coupling constants are likely to be different, and instead of a triplet, a doublet of doublets will be seen. Similarly, if a proton is coupled to two other protons of one type, and a third of another type with a different, smaller coupling constant, then a triplet of doublets is seen. In the example below, the triplet coupling constant is larger than the doublet one. By convention the pattern created by the largest coupling constant is indicated first and the splitting patterns of smaller constants are named in turn. In the case below it would be erroneous to refer to the quartet of triplets as a triplet of quartets. The analysis of such multiplets (which can be much more complicated than the ones shown here) provides important clues to the structure of the molecule being studied.

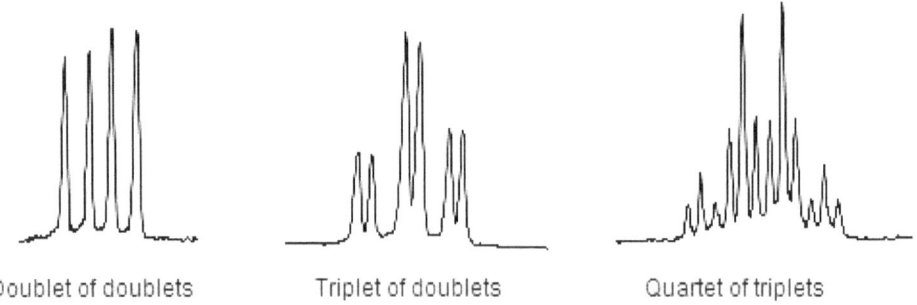

Doublet of doublets   Triplet of doublets   Quartet of triplets

The simple rules for the spin-spin splitting of NMR signals described above apply only if the chemical shifts of the coupling partners are substantially larger than the coupling constant between them. Otherwise there may be more peaks, and the intensities of the individual peaks will be distorted (second-order effects).

13.4 Carbon satellites and spinning sidebands

Occasionally, small peaks can be seen shouldering the main ^1H NMR peaks. These peaks are not the result of proton-proton coupling, but result from the coupling of ^1H atoms to an adjoining carbon-13 (^{13}C) atom. These small peaks are known as carbon satellites as they are small and appear around the main ^1H peak i.e. satellite (around) to them. Carbon satellites are small because only very few of the molecules in the sample have that carbon as the rare NMR-active ^{13}C isotope. As always for coupling due to a single spin-1/2 nucleus, the signal splitting for the H attached to the ^{13}C is a doublet. The H attached to the more abundant ^{12}C is not split, so it is a large singlet. The net result is a pair of evenly-spaced small signals around the main one. If the H signal would already be split due to H–H coupling or other effects, each of the satellites would also reflect this coupling as well (as usual for complex splitting patterns due to dissimilar coupling partners). Other NMR-active nuclei can also cause these satellites, but carbon is most common culprit in the proton NMR spectra of organic compounds.

Sometimes other peaks can be seen around ^{1}H peaks, known as spinning sidebands and are related to the rate of spin of an NMR tube. These are experimental artifacts from the spectroscopic analysis itself, not an intrinsic feature of the spectrum of the chemical and not even specifically related to the chemical or its structure.

Carbon satellites and spinning sidebands should not be confused with impurity peaks.[5]

13.5   See also

- Mass spectrometry

- Pople Notation – letter designations for coupled spin-systems

13.6   References

[1] R. M. Silverstein, G. C. Bassler and T. C. Morrill, *Spectrometric Identification of Organic Compounds*, 5th Ed., Wiley, **1991**.

[2] The Theory of NMR - Chemical Shift

[3] US patent 4110681, Donald C. Hofer; Vincent N. Kahwaty; Carl R. Kahwaty, "NMR field frequency lock system", issued 1978-08-29

[4] Balci, M., in Basci 1H- and 13C-NMR spectroscopy (1st Edition, Elsevier), ISBN 978-0444518118.

[5] Gottlieb HE; Kotlyar V; Nudelman A (October 1997). "NMR Chemical Shifts of Common Laboratory Solvents as Trace Impurities". *J. Org. Chem.* **62** (21): 7512–7515. doi:10.1021/jo971176v. PMID 11671879.

13.7   External links

- ^{1}H-NMR Interpretation Tutorial

- Spectral Database for Organic Compounds

- Proton Chemical Shifts

- 1D Proton NMR 1D NMR experiment

Chapter 14

Proton–proton chain reaction

The **proton–proton chain reaction** is one of two nuclear fusion reactions, along with the CNO cycle, by which stars convert hydrogen to helium and which dominates in stars the size of the Sun or smaller.[1]

In general, proton–proton fusion can occur only if the temperature (i.e. kinetic energy) of the protons is high enough to overcome their mutual electrostatic or Coulomb repulsion.[2]

In the Sun, deuterium-producing events are rare as diprotons, the much more common result of nuclear reactions within the star, immediately decay back into two protons. A complete conversion of the hydrogen in the solar core is calculated to take more than 10^{10} (ten billion) years.[3]

14.1 History of the theory

The theory that proton–proton reactions are the basic principle by which the Sun and other stars burn was advocated by Arthur Stanley Eddington in the 1920s. At the time, the temperature of the Sun was considered too low to overcome the Coulomb barrier. After the development of quantum mechanics, it was discovered that tunneling of the wavefunctions of the protons through the repulsive barrier allows for fusion at a lower temperature than the classical prediction.

Even so, it was unclear how proton–proton fusion might proceed, because the most obvious product, helium-2 (diproton), is unstable and immediately dissociates back into a pair of protons. In 1939, Hans Bethe proposed that one of the protons could beta decay into a neutron via the weak interaction during the brief moment of fusion, making deuterium the initial product in the chain.[4] This idea was part of the body of work in stellar nucleosynthesis for which Bethe won the 1967 Nobel Prize in Physics.

14.2 The pp chain reaction

The first step involves the fusion of two 1H nuclei (protons) into deuterium, releasing a positron and a neutrino as one proton changes into a neutron. It is a two-stage process; first, two protons fuse to form a diproton:

followed by the beta-plus decay of the diproton to deuterium:

with the overall formula:

This first step is extremely slow because the beta-plus decay of the diproton to deuterium is extremely rare (the vast majority of the time, the diproton decays back into hydrogen-1 through proton emission). The half-life of a proton in the core of the Sun before it is involved in a successful p-p fusion is estimated to be a billion years, even at the extreme density and temperatures found there.

The positron emitted by the beta-decay is almost immediately annihilated with an electron, and their mass energy, as well as their kinetic energy, is carried off by two gamma ray photons.

After it is formed, the deuterium produced in the first stage can fuse with another proton to produce a light isotope of helium, 3He:

This process, mediated by the strong nuclear force rather than the weak force, is extremely fast by comparison to the first step. It is estimated that, under the conditions in the Sun's core, a newly created deuterium nucleus exists for only about 4 seconds before it is converted to He-3.

From here there are four possible paths to generate 4He. In pp I, helium-4 is produced by fusing two helium-3 nuclei; the pp II and pp III branches fuse 3He with pre-existing 4He to form beryllium−7, which undergoes further reactions to produce two helium-4 nuclei. In the Sun, the helium-3 produced in these reactions exists for only about 400 years before it is converted into helium-4.[5]

In the Sun, 4He synthesis via branch pp I occurs with a frequency of 86%, pp II with 14% and pp III with 0.11%. There is also an extremely rare pp IV branch. Additionally, other even less frequent reactions may occur; however, the rate of these reactions is very low due to very small cross-sections, or because the number of reacting particles is so low that any reactions that might happen are statistically insignificant. This is partly why no mass-5 or mass-8 elements are seen. While the reactions that would produce them, such as a proton + helium-4 producing lithium-5, or two helium-4 nuclei coming together to form beryllium-8, may *actually* happen, these elements are not detected because there are no stable isotopes of mass 5 or 8; the resulting products immediately decay into their initial reactants.

14.2.1   The pp I branch

The complete pp I chain reaction releases a net energy of 26.22 MeV. Two percent of this energy is lost to the neutrinos that are produced.[6] The pp I branch is dominant at temperatures of 10 to 14 MK. Below 10 MK, the PP chain does not produce much 4He.

14.2.2   The pp II branch

See also: lithium burning

The pp II branch is dominant at temperatures of 14 to 23 MK.

Note that the energies in the equation above are not the energy released by the reaction. Rather, they are the energies of the neutrinos that are produced by the reaction. 90% of the neutrinos produced in the reaction of 7Be to 7Li carry an energy of 0.861 MeV, while the remaining 10% carry 0.383 MeV. The difference is whether the lithium-7 produced is in the ground state or an excited state, respectively.

14.2.3　The pp III branch

The pp III chain is dominant if the temperature exceeds 23 MK.

The pp III chain is not a major source of energy in the Sun (only 0.11%), but was very important in the solar neutrino problem because it generates very high energy neutrinos (up to 14.06 MeV).

14.2.4　The pp IV (Hep) branch

This reaction is predicted but has never been observed due to its rarity (about 0.3 ppm in the Sun). In this reaction, Helium-3 reacts directly with a proton to give helium-4, with an even higher possible neutrino energy (up to 18.8 MeV).

14.2.5　Energy release

Comparing the mass of the final helium-4 atom with the masses of the four protons reveals that 0.007 or 0.7% of the mass of the original protons has been lost. This mass has been converted into energy, in the form of gamma rays and neutrinos released during each of the individual reactions. The total energy yield of one whole chain is 26.73 MeV.

Energy released as gamma rays will interact with electrons and protons and heat the interior of the Sun. Also kinetic energy of fusion products (e.g. of the two protons and the 4
2He from pp-I reaction) increases the temperature of plasma in the Sun. This heating supports the Sun and prevents it from collapsing under its own weight.

Neutrinos do not interact significantly with matter and therefore do not help support the Sun against gravitational collapse. Their energy is lost: the neutrinos in the ppI, ppII and ppIII chains carry away 2.0%, 4.0%, and 28.3% of the energy in those reactions, respectively.[7]

14.3　The pep reaction

Deuterium can also be produced by the rare pep (proton–electron–proton) reaction (electron capture):

In the Sun, the frequency ratio of the pep reaction versus the pp reaction is 1:400. However, the neutrinos released by the pep reaction are far more energetic: while neutrinos produced in the first step of the pp reaction range in energy up to 0.42 MeV, the pep reaction produces sharp-energy-line neutrinos of 1.44 MeV. Detection of solar neutrinos from this reaction were reported by the Borexino collaboration in 2012.[8]

Both the pep and pp reactions can be seen as two different Feynman representations of the same basic interaction, where the electron passes to the right side of the reaction as an anti-electron. This is represented in the figure of proton–proton and electron-capture chain reactions in a star, available at the NDM'06 web site.[9]

14.4　See also

- Triple-alpha process

- CNO cycle

14.5   References

[1] The Proton-Proton Chain

[2] Ishfaq Ahmad, *The Nucleus*, **1**:42,59, (1971), The Proton type-nuclear fission reaction

[3] Kenneth S. Krane, *Introductory Nuclear Physics* , Wiley , 1987, p. 537.

[4] Hans A. Bethe, *Physical Review* **55**:103, 434 (1939); cited in Donald D. Clayton, *Principles of Stellar Evolution and Nucleosynthesis*, The University of Chicago Press, 1983, p. 366.

[5] This time and the two other times above come from: Byrne, J. *Neutrons, Nuclei, and Matter*, Dover Publications, Mineola, New York, 2011, ISBN 0486482383, p 8.

[6] Burbidge, E.; Burbidge, G.; Fowler, William; Hoyle, F. (1 October 1957). "Synthesis of the Elements in Stars". *Reviews of Modern Physics* **29** (4): 547–650. Bibcode:1957RvMP...29..547B. doi:10.1103/RevModPhys.29.547. This value excludes the 2% neutrino energy loss.

[7] Claus E. Rolfs and William S. Rodney, *Cauldrons in the Cosmos*, The University of Chicago Press, 1988, p. 354.

[8] "First Evidence of pep Solar Neutrinos by Direct Detection in Borexino" (preprint on arXiv): Phys. Rev. Lett. **108**, (5), 051302 (2012)

[9] Int'l Conference on Neutrino and Dark Matter, Thursday 07 Sept 2006, http://indico.lal.in2p3.fr/getFile.py/access?contribId=s16t1&sessionId=s16&resId=1&materialId=0&confId=a05162 Session 14.

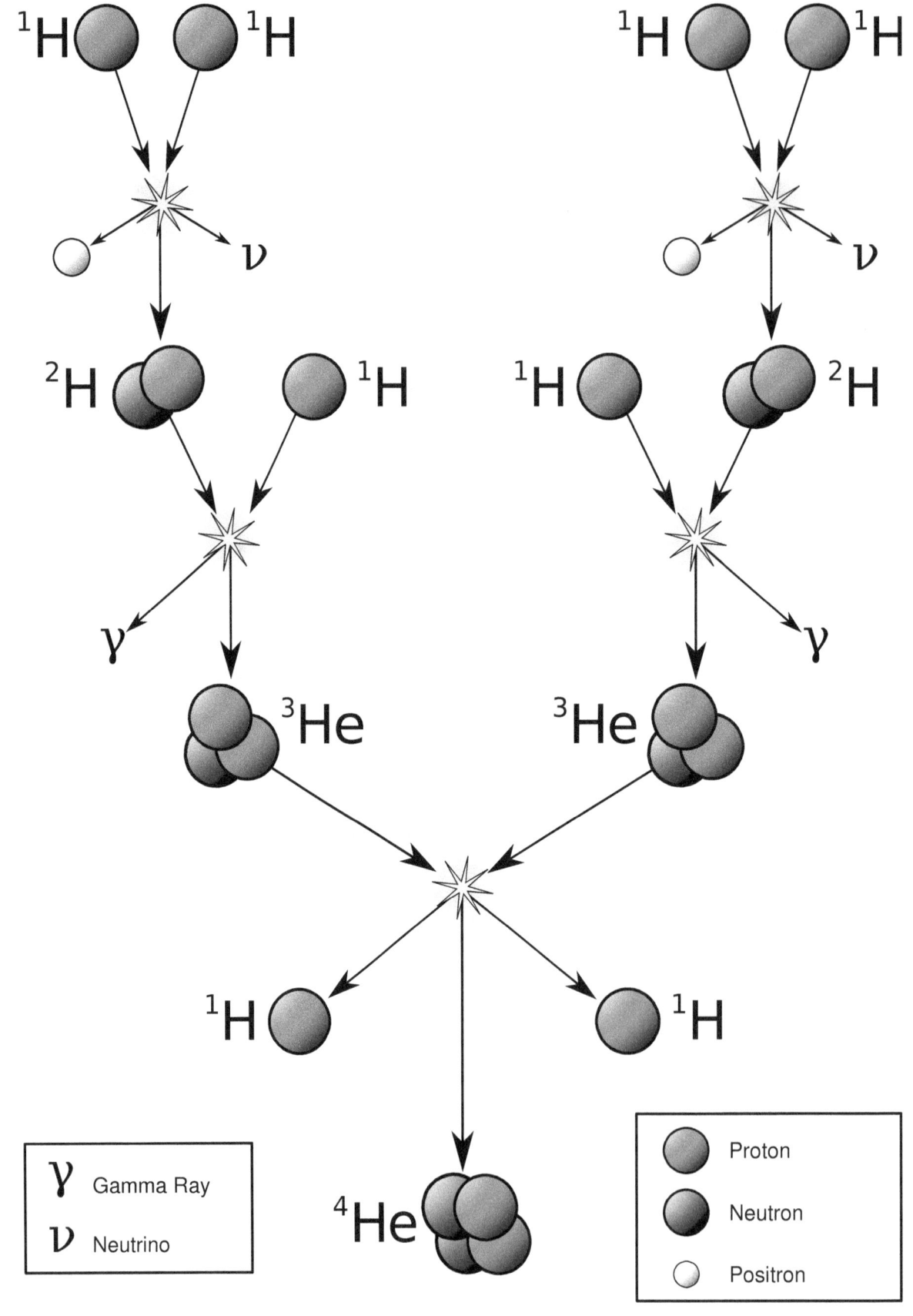

The proton–proton chain reaction dominates in stars the size of the Sun or smaller.

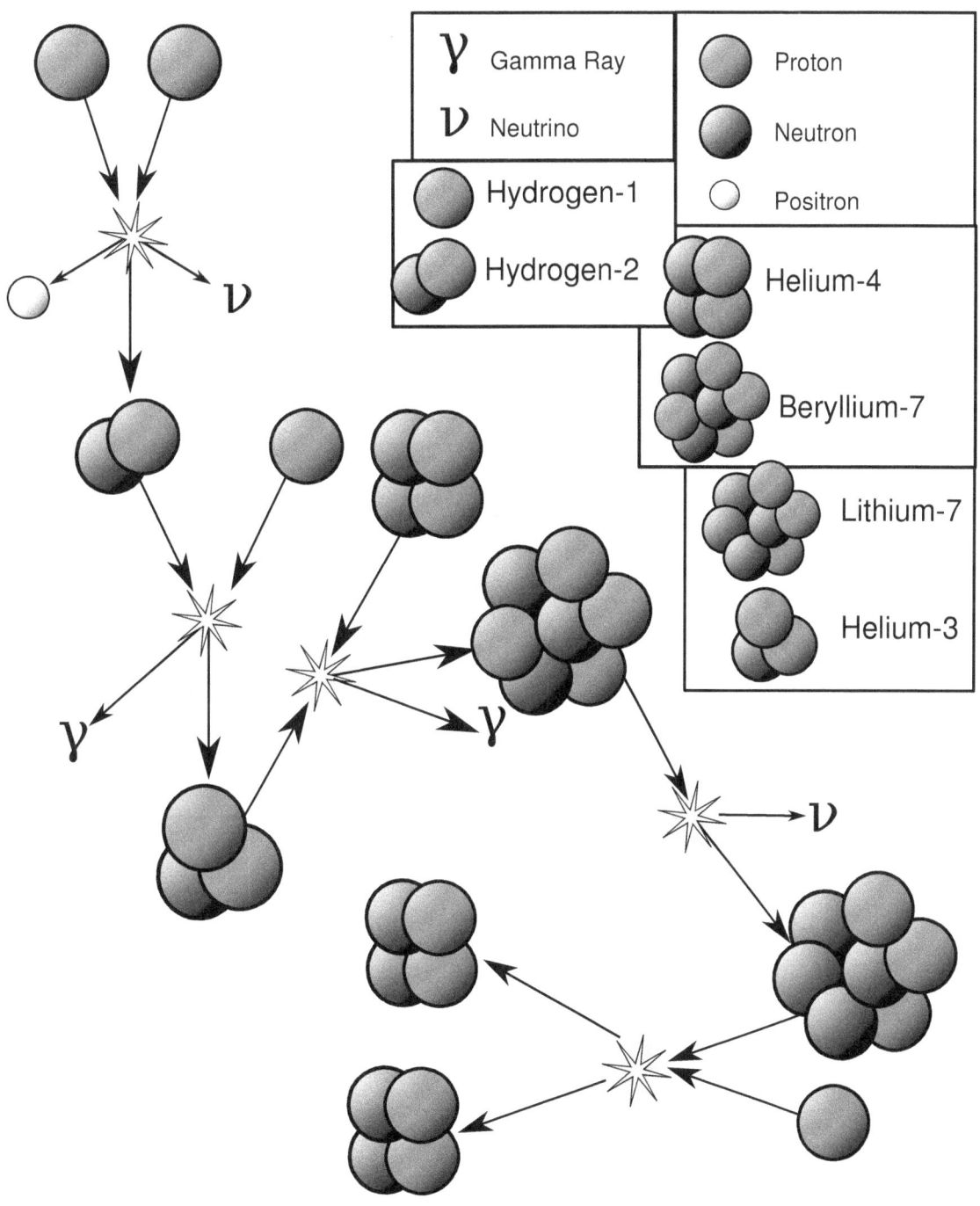

Proton–proton II chain reaction

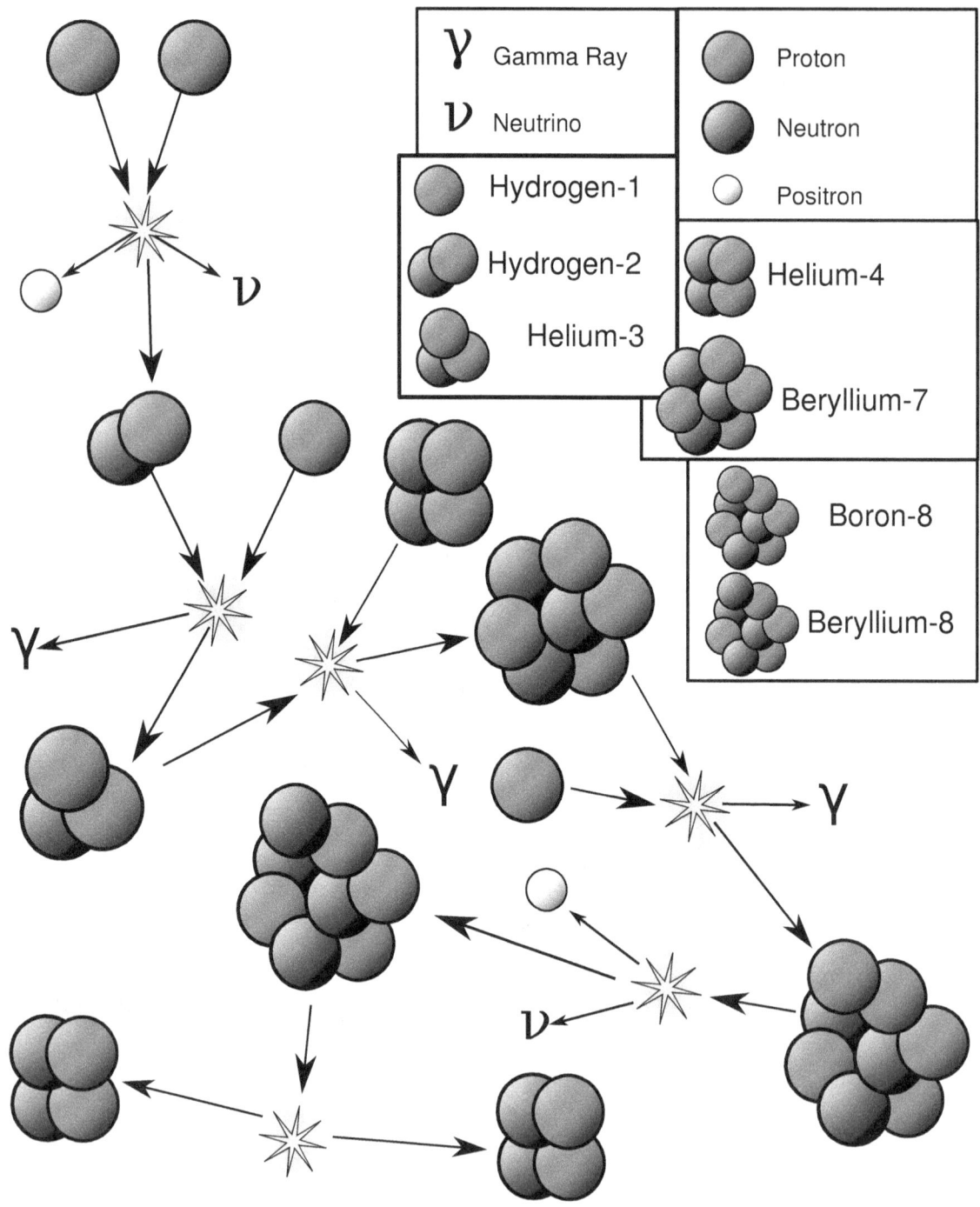

Proton–proton III chain reaction

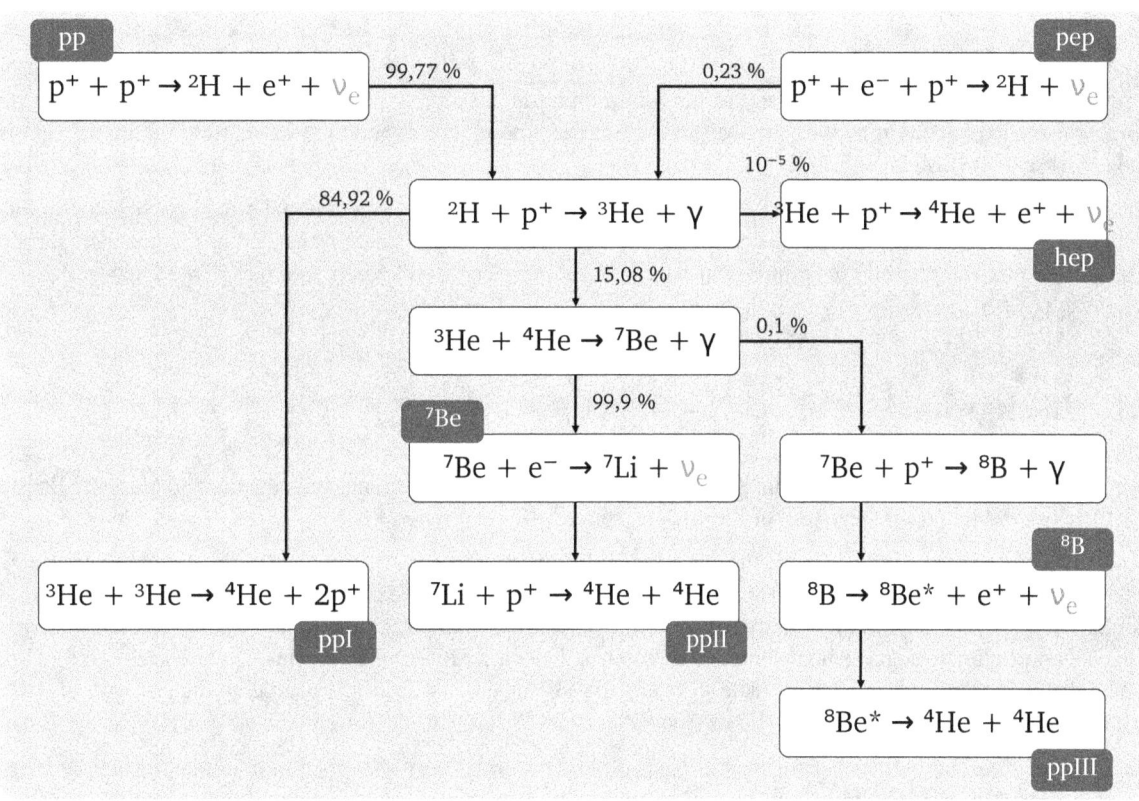

Proton–proton and electron-capture chain reactions in a star.

Chapter 15

Proton spin crisis

The **proton spin crisis** (sometimes called the "proton spin puzzle") is a theoretical crisis precipitated by an experiment in 1987 which tried to determine the spin configuration of the proton. The experiment was carried out by the European Muon Collaboration (EMC).[1]

Physicists expected that the quarks carry all the proton spin. However, not only was the total proton spin carried by quarks far smaller than 100%, these results were consistent with almost zero proton spin being carried by quarks. This surprising and puzzling result was termed the "proton spin crisis".[2] The problem is still considered one of the important unsolved problems in physics.

15.1 Background

A key question is how the nucleon's spin is distributed amongst its constituent partons (quarks and gluons). Physicists originally expected that quarks carry all of the nucleon spin.

According to quantum chromodynamics, the proton is built from two *up* and one *down* quark, gluons and possibly additional pairs of quark and anti-quark.[3] The ruling assumption was that since the proton is stable, then it exists in the lowest possible energy level. Therefore, it was expected that the quark's wave function is the spherically symmetric s-wave with no spatial contribution to angular momentum. The proton is, like each of its quarks, a spin-1/2 particle. Therefore, it was assumed that two of the quarks have opposite spins and the spin of the third quark is parallel to the proton spin.

15.2 The experiment

In this EMC experiment, a quark of a polarized proton target was hit by a polarized muon beam, and the quark's instantaneous spin was measured. In a polarized proton target, all the protons' spin take the same direction, and therefore it was expected that the spin of two out of the three quarks cancels out and the spin of the third quark is polarized in the direction of the proton's spin. Thus, the sum of the quarks' spin was expected to be equal to the proton's spin.

However, it was found in this EMC experiment that the number of quarks with spin in the proton's spin direction was almost the same as the number of quarks whose spin was in the opposite direction. This is the proton spin crisis. Similar results have been obtained in later experiments.[4]

15.3 Recent work

A 2008 work shows that more than half of the spin of the proton stems from the motion of its quarks, and the missing spin is produced by the quarks' spatial angular momentum.[5] This work uses relativistic effects together with other QCD properties and explains how they boil down to an overall spatial angular momentum that is consistent with the experimental data. 2014 Results from the RHIC are beginning to measure the gluon contribution, and are working to increase precision.

Results from recent experiments reveal new insights about how quarks and gluons, the subatomic building blocks of matter, contribute to proton "spin".[6]

15.4   References

[1] Ashman, J.; EMC Collaboration (1988). "A measurement of the spin asymmetry and determination of the structure function g1 in deep inelastic muon-proton scattering". *Physics Letters B* **206** (2): 364. Bibcode:1988PhLB..206..364A. doi:10.1016/0370-2693(88)91523-7.

[2] Londergan, J. T. (2009). "Nucleon Resonances and Quark Structure". *International Journal of Modern Physics E* **18** (5–6): 1135. arXiv:0907.3431. Bibcode:2009IJMPE..18.1135L. doi:10.1142/S0218301309013415.

[3] Halzen, F.; Martin, A. D. (1984). *Quarks and Leptons: An Introductory Course in Modern Particle Physics*. Wiley. ISBN 978-0-471-88741-6.

[4] Jaffe, R. (1995). "Where does the proton really get its spin?" (PDF). *Physics Today* **48** (9): 24–30. Bibcode:1995PhT....48i..24J. doi:10.1063/1.881473.

[5] Thomas, A. (2008). "Interplay of Spin and Orbital Angular Momentum in the Proton". *Physical Review Letters* **101** (10): 102003. arXiv:0803.2775. Bibcode:2008PhRvL.101j2003T. doi:10.1103/PhysRevLett.101.102003.

[6] "Physicists narrow search for solution to proton spin puzzle", *Brookhaven National Laboratory*, November 4, 2014.

15.5   External links

- http://phys.org/news/2013-04-quarks-dictate-proton.html

Chapter 16

Antiproton

The **antiproton**, p, (pronounced *p-bar*) is the antiparticle of the proton. Antiprotons are stable, but they are typically short-lived since any collision with a proton will cause both particles to be annihilated in a burst of energy.

The existence of the antiproton with −1 electric charge, opposite to the +1 electric charge of the proton, was predicted by Paul Dirac in his 1933 Nobel Prize lecture.[1] Dirac received the Nobel Prize for his previous 1928 publication of his Dirac Equation that predicted the existence of positive and negative solutions to the Energy Equation ($E = mc^2$) of Einstein and the existence of the positron, the antimatter analog to the electron, with positive charge and opposite spin.

The antiproton was experimentally confirmed in 1955 by University of California, Berkeley physicists Emilio Segrè and Owen Chamberlain, for which they were awarded the 1959 Nobel Prize in Physics. An antiproton consists of two up antiquarks and one down antiquark (uud). The properties of the antiproton that have been measured all match the corresponding properties of the proton, with the exception that the antiproton has electric charge and magnetic moment that are the opposites of those in the proton. The questions of how matter is different from antimatter, and the relevance of antimatter in explaining how our universe survived the Big Bang remain open problems—open, in part, due to the relative dearth of antimatter in today's universe.

16.1   Occurrence in nature

Antiprotons have been detected in cosmic rays for over 25 years, first by balloon-borne experiments and more recently by satellite-based detectors. The standard picture for their presence in cosmic rays is that they are produced in collisions of cosmic ray protons with nuclei in the interstellar medium, via the reaction, where A represents a nucleus:

p + A → p+ p +p+ A

The secondary antiprotons (p) then propagate through the galaxy, confined by the galactic magnetic fields. Their energy spectrum is modified by collisions with other atoms in the interstellar medium, and antiprotons can also be lost by "leaking out" of the galaxy.

The antiproton cosmic ray energy spectrum is now measured reliably and is consistent with this standard picture of antiproton production by cosmic ray collisions.[2] This sets upper limits on the number of antiprotons that could be produced in exotic ways, such as from annihilation of supersymmetric dark matter particles in the galaxy or from the evaporation of primordial black holes. This also provides a lower limit on the antiproton lifetime of about 1-10 million years. Since the galactic storage time of antiprotons is about 10 million years, an intrinsic decay lifetime would modify the galactic residence time and distort the spectrum of cosmic ray antiprotons. This is significantly more stringent than the best laboratory measurements of the antiproton lifetime:

- LEAR collaboration at CERN: 0.08 years

- Antihydrogen Penning trap of Gabrielse et al.: 0.28 years[3]

127

- APEX collaboration at Fermilab: 50000 years for p → μ− + anything

- APEX collaboration at Fermilab: 300000 years for p → e− + γ

The magnitude of properties of the antiproton are predicted by CPT symmetry to be exactly related to those of the proton. In particular, CPT symmetry predicts the mass and lifetime of the antiproton to be the same as those of the proton, and the electric charge and magnetic moment of the antiproton to be opposite in sign and equal in magnitude to those of the proton. CPT symmetry is a basic consequence of quantum field theory and no violations of it have ever been detected.

16.1.1   List of recent antiproton cosmic ray detection experiments

- BESS: balloon-borne experiment, flown in 1993, 1995, 1997, 2000, 2002, 2004 (Polar-I) and 2007 (Polar-II).

- CAPRICE: balloon-borne experiment, flown in 1994[4] and 1998.

- HEAT: balloon-borne experiment, flown in 2000.

- AMS: space-based experiment, prototype flown on the space shuttle in 1998, intended for the International Space Station, launched May 2011.

- PAMELA: satellite experiment to detect cosmic rays and antimatter from space, launched June 2006. Recent report discovered 28 antiprotons in the South Atlantic Anomaly.[5]

16.2   Modern experiments and applications

Antiproton accumulator (center) at Femilab[6]

Antiprotons are routinely produced at Fermilab for collider physics operations in the Tevatron, where they are collided with protons. The use of antiprotons allows for a higher average energy of collisions between quarks and antiquarks than would be possible in proton-proton collisions. This is because the valence quarks in the proton, and the valence antiquarks in the antiproton, tend to carry the largest fraction of the proton or antiproton's momentum.

Their formation requires energy equivalent to a temperature of 10 trillion K (10^{13} K) and this does not tend to happen naturally. However, at CERN, protons are accelerated in the Proton Synchrotron to an energy of 26 GeV, and then smashed into an iridium rod. The protons bounce off the iridium nuclei with enough energy for matter to be created. A range of particles and antiparticles are formed, and the antiprotons are separated off using magnets in vacuum.

In July 2011, the ASACUSA experiment at CERN determined the mass of the antiproton to be 1836.1526736(23) times more massive than an electron.[7] This is the same as the mass of a proton, within the level of certainty of the experiment.

Antiprotons have been shown within laboratory experiments to have the potential to treat certain cancers, in a similar method currently used for ion (proton) therapy.[8] The primary difference between antiproton therapy and proton therapy is that following ion energy deposition the antiproton annihilates depositing additional energy in the cancerous region.

16.3 See also

- Antimatter

- Antineutron

- Positron

- Antihydrogen

- Antiprotonic helium

- List of particles

- Recycling antimatter

16.4 References

[1] Dirac, Paul A. M. (1933). "Theory of electrons and positrons" (PDF)

[2] Kennedy, Dallas C. (2000). "Cosmic Ray Antiprotons". *Proc. SPIE*. Gamma-Ray and Cosmic-Ray Detectors, Techniques, and Missions **2806**: 113. arXiv:astro-ph/0003485. doi:10.1117/12.253971.

[3] Caso, C. et al. (1998). "Particle Data Group" (PDF). *European Physical Journal C* **3**: 613. Bibcode:1998EPJC....3....1P. doi:10.1007/s10052-998-0104-x.

[4] Caprice Experiment

[5] Adriani, O.; Barbarino, G. C.; Bazilevskaya, G. A.; Bellotti, R.; Boezio, M.; Bogomolov, E. A.; Bongi, M.; Bonvicini, V.; Borisov, S.; Bottai, S.; Bruno, A.; Cafagna, F.; Campana, D.; Carbone, R.; Carlson, P.; Casolino, M.; Castellini, G.; Consiglio, L.; De Pascale, M. P.; De Santis, C.; De Simone, N.; Di Felice, V.; Galper, A. M.; Gillard, W.; Grishantseva, L.; Jerse, G.; Karelin, A. V.; Kheymits, M. D.; Koldashov, S. V. et al. (2011). "The Discovery of Geomagnetically Trapped Cosmic-Ray Antiprotons". *The Astrophysical Journal Letters* **737** (2): L29. arXiv:1107.4882v1. Bibcode:2011ApJ...737L..29A. doi:10.1088/2041-8205/737/2/L29.

[6] Nagaslaev, V. (17 May 2007). *Antiproton Production at Fermilab* (PDF). Retrieved 14 August 2015.

[7] Hori, M.; Barna, Daniel; Dax, Andreas; Hayano, Ryugo; Friedreich, Susanne; Juhász, Bertalan; Pask, Thomas et al. (2011). "Two-photon laser spectroscopy of antiprotonic helium and the antiproton-to-electron mass ratio". *Nature* **475** (7357): 484–8. doi:10.1038/nature10260. PMID 21796208.

[8] "Antiproton portable traps and medical applications" (PDF).

16.5 Text and image sources, contributors, and licenses

16.5.1 Text

- **Proton** *Source:* https://en.wikipedia.org/wiki/Proton?oldid=682675320 *Contributors:* Mav, Bryan Derksen, Zundark, Manning Bartlett, Ap, Andre Engels, Josh Grosse, Danny, XJaM, Ellmist, Heron, Jaknouse, Stevertigo, Dwmyers, Bdesham, Patrick, Kku, Stewacide, TakuyaMurata, Egil, NuclearWinner, Looxix~enwiki, Mkweise, Ahoerstemeier, Sobekhotep, Salsa Shark, Glenn, Kaihsu, Jordi Burguet Castell, Mxn, Denny, Timwi, Paul-L~enwiki, Omegatron, Geraki, David.Monniaux, Donarreiskoffer, Robbot, Josh Cherry, Jakohn, RedWolf, Altenmann, Yelyos, Merovingian, Flauto Dolce, Hadal, Giftlite, Mikez, Tom harrison, Lupin, Herbee, Xerxes314, Fleminra, Bensaccount, Foobar, PlatinumX, Utcursch, Knutux, Antandrus, Beland, Eroica, Melikamp, Rdsmith4, DragonflySixtyseven, Icairns, Cglassey, Deglr6328, Grunt, Thorwald, Jenlight, Mike Rosoft, Diagonalfish, Discospinster, Cacycle, Vsmith, Jpk, Wikiacc, Mani1, Bender235, Kjoonlee, RJHall, El C, Shrike, Femto, Bobo192, O18, Army1987, Smalljim, GTubio, Vortexrealm, Elipongo, Foobaz, Kjkolb, Obradovic Goran, Nsaa, Eddideigel, Anthony Appleyard, Mattpickman, Apoc2400, Carmelbuck, Spangineer, Wtmitchell, Saga City, Uucp, Crobzub, Vcelloho, RainbowOfLight, TenOfAllTrades, Computerjoe, Kusma, Itsmine, Falcorian, Richard Arthur Norton (1958-), Firien, GregorB, Macaddct1984, Mayz, Karam.Anthony.K, Marudubshinki, Bebenko, Rtcpenguin, Graham87, Kbdank71, Ketiltrout, Drbogdan, Rjwilmsi, Strait, NeonMerlin, RadicalJester, Bubba73, Yamamoto Ichiro, Rangek, FlaBot, Nivix, RexNL, Fresheneesz, Srleffler, Imnotminkus, King of Hearts, CiaPan, Chobot, Deyyaz, Roboto de Ajvol, The Rambling Man, YurikBot, Bambaiah, JWB, Anuran, Pip2andahalf, RussBot, Wigie, Jumbo Snails, Raquel Baranow, Hellbus, Salsb, Oni Lukos, Anomalocaris, NawlinWiki, Injinera, Welsh, Długosz, Martin Ulfvik, Moe Epsilon, BOT-Superzerocool, DeadEyeArrow, Bota47, D-Day, Mtu, Pooryorick~enwiki, J S Ayer, Theodolite, Bayerischermann, Closedmouth, Reyk, Petri Krohn, DGaw, Paul D. Anderson, Katieh5584, RG2, SDS, GrinBot~enwiki, Nekura, Orii, Luk, Sycthos, Itub, Attilios, SmackBot, Moeron, Incnis Mrsi, Melchoir, CyclePat, Edgar181, HalfShadow, Dhochron, Munky2, Gilliam, Chris the speller, Rajeevmass~enwiki, Rkitko, AndrewBuck, Bethling, SchfiftyThree, Complexica, Sbharris, Rogermw, Can't sleep, clown will eat me, Shalom Yechiel, PeteShanosky, Writtenright, Homestarmy, Wikiwikiwiki3~enwiki, SundarBot, COMPFUNK2, Dreadstar, Orczar, Drphilharmonic, Dvorak729, DMacks, Mion, Vina-iwbot~enwiki, Bdushaw, SashatoBot, ArglebargleIV, Khazar, Cholerashot, Rijkbenik, Spacecadethailey, Herr apa, Deathcakes, Noah Salzman, Aeluwas, Waggers, Mozzura, Mattabat, Elb2000, Newone, MOBle, Igoldste, Rhetth, Frank Lofaro Jr., Tawkerbot2, CmdrObot, Ale jrb, Sir Vicious, KyraVixen, Ruslik0, McVities, TheTito, Cydebot, Nick Y., Gogo Dodo, Red Director, Umdunno, Difluoroethene, Odie5533, Q43, Tawkerbot4, Dwool99f, Narayanese, Rasheedy, Zalgo, Lo2u, Jenswort, Thijs!bot, Epbr123, Tsogo3, Headbomb, Marek69, Electron9, Mnemeson, Dfrg.msc, Philippe, Aadal, AntiVandalBot, Seaphoto, HairyDan, Shirt58, EarthPerson, Gregnx, Jj137, Naturalnumber, Myanw, Ellissound, Leuko, MER-C, CosineKitty, Fetchcomms, Andonic, Kerotan, .anacondabot, Acroterion, Plynn9, Casmith_789, Bongwarrior, VoABot II, Astrangequark, Swpb, WODUP, Recurring dreams, Avicennasis, Catgut, Dirac66, 28421u2232nfenfcenc, Hveziris, Fang 23, The Real Marauder, Oddworth, JaGa, MartinBot, Rettetast, Pbroks13, Artaxiad, J.delanoy, WeglarczykJ, Silverxxx, C.A.T.S. CEO, Maurice Carbonaro, 12dstring, WarthogDemon, Acalamari, Exdejesus, TomasBat, Antony-22, Potatoswatter, KylieTastic, Vanished user 39948282, Treisijs, S, SoCalSuperEagle, Xiahou, Specter01010, Idioma-bot, Ciju, 28bytes, VolkovBot, Doc7777777777, Jeff G., Soliloquial, Tuffcarrot, Philip Trueman, TXiKiBoT, The Original Wildbear, Vipinhari, Bjman, Bigyaks, Alexalexalex123~enwiki, Meters, Antixt, Spinningspark, Insanity Incarnate, Upquark, AlleborgoBot, Vitalikk, B41988, Petergans, Demmy100, SieBot, Accounting4Taste, Jauerback, Studnic12, Xe1881, Yintan, GlassCobra, Keilana, RadicalOne, Flyer22, Sbowers3, Prestonmag, Oxymoron83, BenoniBot~enwiki, Jacob.jose, Mygerardromance, Rajbboy69, ClueBot, The Thing That Should Not Be, RODERICKMOLASAR, Tigerboy1966, Regibox, ChandlerMapBot, Mr blabla, Excirial, Alexbot, Robbie098, Poopmister91191, Ploft, NuclearWarfare, Lunchscale, PhySusie, SoxBot, El pobre Pedro, Thehelpfulone, La Pianista, Thingg, Kanxkawii, Aitias, Subash.chandran007, Johnuniq, XLinkBot, Avoided, WikHead, SilvonenBot, SkyLined, Mls1492, Weletahoozyzog, Addbot, Zrules, Arcturus87, Ronhjones, Cst17, Glane23, AndersBot, Wandering Traveler, Omnipedian, LinkFA-Bot, Numbo3-bot, Tide rolls, Thermalimage, Luckas-bot, Yobot, Велетень, Wickedwizardofoz, Newportm, Kilom691, Heart of a Lion, Eric-Wester, Jay0205, AnomieBOT, LeftyAce, Götz, Jim1138, Sp eloc, Judoc, Materialscientist, The High Fin Sperm Whale, Citation bot, OllieFury, Apollo, Xqbot, Phazvmk, Blennow, Cureden, Wyklety, Aa77zz, Squishywushy123, Srich32977, Rueyfgugdtj, RibotBOT, PM800, A. di M., ⁇⁇, Rain bowell, CES1596, FrescoBot, Paine Ellsworth, Tobby72, Citation bot 1, Pinethicket, HRoestBot, Calmer Waters, Bejinhan, Impala2009, Nicklcms, Blckmgc, SkyMachine, Gryllida, Double sharp, TobeBot, Ilovefatchicks, Keegscee, DARTH SIDIOUS 2, TjBot, StudentDoc73, Nachos0123, EmausBot, WikitanvirBot, ANDREVV, Bencbartlett, Zues zeus kratos, Pcorty, Sterrettc, K6ka, JSquish, Stuffness12, John Cline, Harddk, Fæ, StringTheory11, Brazmyth, Suslindisambiguator, Wayne Slam, Zach444, Rcsprinter123, Wiggles007, Brandmeister, Donner60, Sarthak 94, Xonqnopp, ClueBot NG, Timelord360, This lousy T-shirt, IHopeThisNameIsntTaken, Corusant, Cntras, Dictabeard, Rezabot, Helpful Pixie Bot, Wbm1058, Bibcode Bot, BG19bot, ArthropodOfDoom, RadioActiveKitKat, AvocatoBot, Metricopolus, JacobTrue, Mlkamitso, Toccata quarta, Blaspie55, Mhutchison43, 220 of Borg, Anbu121, Hitheresir, RudolfRed, Knodir, BattyBot, ChrisGualtieri, Hower64, Ducknish, Stephen Glass, BrightStarSky, Dexbot, Mogism, Sheehan Cein14, Frosty, Pidotclan, Marcoapc.84, Delnium strex, Faizan, Huddydakota, Prof.Professer, Dustin V. S., Borreswafflertron, Ugog Nizdast, Jwratner1, Javierha, JaconaFrere, AspaasBekkelund, Bballbro62, Melcous, Monkbot, Jayakumar RG, Scorpion1045, Krebs49, Yollowswagger19, Maddie005, Junchuann, Chrisbrownthathoe, Orgasam069, Tktobykerby, Interpuncts, Tetra quark, Cjohnson2020, KasparBot, Muzammil Alam Baig, Rambunctious Racoon, Teo boruch and Anonymous: 638

- **Subatomic particle** *Source:* https://en.wikipedia.org/wiki/Subatomic_particle?oldid=681326892 *Contributors:* The Anome, Tarquin, Michael Hardy, FrankH, Ixfd64, CesarB, NuclearWinner, Looxix~enwiki, Ahoerstemeier, LittleDan, Glenn, Kwekubo, Schneelocke, Bevo, Chrisjj, Donarreiskoffer, Baldhur, Romanm, Anthony, Alan Liefting, Giftlite, DocWatson42, Awolf002, Mintleaf~enwiki, Dissident, Xerxes314, Everyking, Bensaccount, Vadmium, Antandrus, Rdsmith4, Kenny TM~~enwiki, Discospinster, ElTyrant, Vsmith, ESkog, RJHall, El C, Bobo192, Shenme, PiccoloNamek, Stephen G. Brown, Alansohn, Ctande, MarkGallagher, Wtshymanski, Egg, DV8 2XL, Ceyockey, Adrian.benko, Oleg Alexandrov, Mindmatrix, JarlaxleArtemis, Duncan.france, Isnow, SeventyThree, Dysepsion, Rjwilmsi, Strait, Klassykittychick, Scorpiuss, Boccobrock, Erkcan, Naraht, DannyWilde, SouthernNights, King of Hearts, Chobot, Wavelength, Bambaiah, Jimp, Phantomsteve, Loom91, Stephenb, PoorLeno, Bachrach44, Spike Wilbury, Syrthiss, Gat0r, Wknight94, Light current, KGasso, GraemeL, Rlove, Sitenl, Asterion, SmackBot, Incnis Mrsi, JohnRussell, Darkgod, Jordan.ambra, Chris the speller, Bluebot, Persian Poet Gal, DHN-bot~enwiki, Sbharris, Can't sleep, clown will eat me, Drkirkby, Vladis1av, Voyajer, Pax85, Radagast83, Edwtie, Drphilharmonic, Thinkingman, Lambiam, Doug Bell, Rigadoun, Ortho, 041744, Ckatz, RandomCritic, Ginkgo100, Esurnir, Tawkerbot2, JForget, Megaboz, Johnlogic, Myasuda, Christian75, Thijs!bot, Epbr123, Guyla, Mbell, Nonagonal Spider, Headbomb, Mjollnir783, Weasel5i2, Escarbot, Ssr, Mentifisto, AntiVandalBot, Luna Santin, Refried, JAnDbot, Instinct, Acroterion, Bongwarrior, VoABot II, Ling.Nut, Glen, Geboy, Mike6271, Davburns, J.delanoy, Yonide-

Boud, Michael Hardy, SebastianHelm, Looxix~enwiki, Julesd, Glenn, AugPi, Mxn, Raven in Orbit, Reddi, Phr, Tpbradbury, Populus, Haoherb428, Phys, Floydian, Bevo, Pierre Boreal, AnonMoos, BenRG, Jeffq, Dmytro, Drxenocide, Robbot, Nurg, Securiger, Texture, Roscoe x, Fuelbottle, Mattflaschen, Tobias Bergemann, Alan Liefting, Ancheta Wis, Giftlite, Dbenbenn, Harp, Herbee, Monedula, LeYaYa, Xerxes314, Dratman, Alison, JeffBobFrank, Dmmaus, Pharotic, Brockert, Bodhitha, Andycjp, Sonjaaa, HorsePunchKid, APH, Icairns, AmarChandra, Gscshoyru, Kate, Arivero, FT2, Rama, Vsmith, David Schaich, Xezbeth, D-Notice, Dfan, Bender235, Pt, El C, Laurascudder, Shanes, Drhex, Fogger~enwiki, Brim, Rbj, Jeodesic, Jumbuck, Alansohn, Gary, ChristopherWillis, Guy Harris, Axl, Sligocki, Kocio, Stillnotelf, Alinor, Wtmitchell, Egg, TenOfAllTrades, H2g2bob, Killing Vector, Linas, Mindmatrix, Benbest, Dodiad, Mpatel, Faethon, TPickup, Faethon34, Palica, Dysepsion, Faethon36, Qwertyca, Drbogdan, Rjwilmsi, Zbxgscqf, Macumba, Strangethingintheland, Dstudent, R.e.b., Bubba73, Drrngrvy, Agasicles, FlaBot, Naraht, Agasides, DannyWilde, Dave1g, Itinerant1, Gparker, Jrtayloriv, Goudzovski, Chobot, Bgwhite, FrankTobia, YurikBot, Bambaiah, Ohwilleke, VoxMoose, Bhny, JabberWok, Bovineone, Krbabu, SCZenz, JulesH, Davemck, Lomn, E2mb0t~enwiki, Dna-webmaster, Jrf, Dv82matt, Tetracube, Hirak 99, Arthur Rubin, Netrapt, JLaTondre, Caco de vidro, RG2, GrinBot~enwiki, That Guy, From That Show!, Hal peridol, SmackBot, YellowMonkey, Tom Lougheed, Melchoir, Bazza 7, KocjoBot~enwiki, Jagged 85, Thunderboltz, Setanta747 (locked), Skizzik, Dauto, Chris the speller, Bluebot, TimBentley, Sirex98, Silly rabbit, Complexica, Metacomet, DHN-bot~enwiki, MovGP0, QFT, Kittybrewster, Addshore, Jmnbatista, Cybercobra, Jgwacker, BullRangifer, Soarhead77, Daniel.Cardenas, Yevgeny Kats, Byelf2007, TriTertButoxy, Craig Bolon, Ajnosek, Ekjon Lok, Bjankuloski06, Tarcieri, Waggers, JarahE, Michaelbusch, Lottamiata, Newone, Twas Now, IanOfNorwich, Srain, Patrickwooldridge, J Milburn, Mosaffa, Gatortpk, Vessels42, Geremia, Van helsing, Harrigan, Phatom87, Cydebot, David edwards, Verdy p, Michael C Price, Xantharius, Crum375, JamesAM, Thijs!bot, Epbr123, Headbomb, Phy1729, Stannered, Tariqhada, Seaphoto, Orionus, Voyaging, Gnixon, Jbaranao, Jrw@pobox.com, Len Raymond, Narssarssuaq, Bakken, CattleGirl, Davidoaf, Vanished user ty12kl89jq10, Lvwarren, Taborgate, Leyo, HEL, J.delanoy, Hans Dunkelberg, Stephanwehner, Wbellido, Aoosten, Jacksonwalters, The Transliterator, DadaNeem, Student7, Joshmt, WJBscribe, Jozwolf, Hexane2000, BernardZ, Awren, Sheliak, Physicist brazuca, Schucker, Goop Goop, Fences and windows, Dextrose, Mcewan, Swamy g, TXiKiBoT, Sharikkamur, Thrawn562, Voorlandt, Escalona, Setreset, PDFbot, Pleroma, UnitedStatesian, Piyush Sriva, Kacser, Billinghurst, Francis Flinch, Moose-32, Ptrslv72, David Barnard, SieBot, ShiftFn, Robdunst, Jim E. Black, SheepNotGoats, Gerakibot, Nozzer42, Mr swordfish, Wing gundam, Bamkin, Likebox, Arthur Smart, HungarianBarbarian, Commutator, KathrynLybarger, Iomesus, C0nanPayne, Crazz bug 5, ClueBot, Superwj5, Wwheaton, Garyzx, SuperHamster, Elsweyn, Maldmac, DragonBot, Djr32, Diagramma Della Verita, Nymf, Eeekster, Brews ohare, NuclearWarfare, PhySusie, Ordovico, Mastertek, DumZiBoT, BodhisattvaBot, Guarracino, Mitch Ames, Truthnlove, Stephen Poppitt, Tayste, Addbot, Deepmath, Eric Drexler, DWHalliday, Mjamja, Leszek Jańczuk, NjardarBot, Mwoldin, Bassbonerocks, Barak Sh, AgadaUrbanit, Lightbot, Smeagol 17, Abjiklam, Ve744, Luckas-bot, Yobot, Orion11M87, AnomieBOT, JackieBot, Icalanise, Citation bot, ArthurBot, Northryde, LilHelpa, Xqbot, Sionus, Professor J Lawrence, Tomwsulcer, Edsegal, GrouchoBot, Trongphu, QMarion II, Ernsts, A. di M., Bytbox, FrescoBot, Paine Ellsworth, Aliotra, Steve Quinn, Citation bot 1, Rameshngbot, MJ94, RedBot, MastiBot, Aknochel, Sijothankam, Puzl bustr, Beta Orionis, Physics therapist, Bj norge, Innotata, Jesse V., RjwilmsiBot, Mathewsyriac, Afteread, EmausBot, Bookalign, WikitanvirBot, Wilhelm-physiker, Bdijkstra, DerNeedle, Kenmint, Dbraize, Tanner Swett, HeptishHotik, بەهار, ممنشین, Suslindisambiguator, Quondum, Webbeh, UniversumExNihilo, Vanished user fijw983kjaslkekfhj45, Maschen, RockMagnetist, Stormymountain, Ζετα ζ, Whoop whoop pull up, Isocliff, ClueBot NG, Smtchahal, Snotbot, Tonypak, O.Koslowski, CharleyQuinton, Dsperlich, Theopolisme, ZakMarksbury, Helpful Pixie Bot, Bibcode Bot, BG19bot, Tirebiter78, AvocatoBot, Lukys~enwiki, Stapletongrey, Ownedroad9, Chip123456, ChrisGualtieri, Khazar2, Billyfesh399, Rhlozier, JYBot, Dexbot, Doom636, Rongended, Cerabot~enwiki, CuriousMind01, Cjean42, Jayanta mallick, Joeinwiki, Kowtje, JPaestpreornJeolhlna, Eyesnore, Euan Richard, Nigstomper, Particle physicist, Prokaryotes, Jernahthern, Ginsuloft, Dimension10, JNrgbKLM, Krabaey, 1codesterS, FelixRosch, Monkbot, Delbert7, BradNorton1979, Lathamboyle, Tetra quark, KasparBot, Buckbill10 and Anonymous: 357

- **Charge radius** *Source:* https://en.wikipedia.org/wiki/Charge_radius?oldid=678447886 *Contributors:* Eliasen, Rjwilmsi, Physchim62, Jimp, Ohwilleke, GraemeMcRae, Hmains, Ruslik0, Dirac66, Leyo, John Pons, Addbot, Yobot, Bibcode Bot, Dexbot, Anrnusna and Anonymous: 4

- **Proton decay** *Source:* https://en.wikipedia.org/wiki/Proton_decay?oldid=679823382 *Contributors:* Uriyan, Bryan Derksen, Tarquin, Taw, Alex.tan, Maury Markowitz, Stevertigo, Edward, Michael Hardy, Bcrowell, TakuyaMurata, Alfio, Looxix~enwiki, J'raxis, Bluelion, Maximus Rex, Phys, TravelingDude, BenRG, Securiger, Henrygb, Bkell, Wereon, Giftlite, Herbee, Xerxes314, DJSupreme23, CryptoDerk, Cglassey, B.d.mills, Deglr6328, Rich Farmbrough, FT2, Pjacobi, Ben Standeven, El C, I9Q79oL78KiL0QTFHgyc, La goutte de pluie, Jason One, Miranche, Zyqqh, Reaverdrop, DV8 2XL, Feezo, MattJakel, Joke137, Christopher Thomas, Rjwilmsi, RexNL, Jimp, Limulus, Jengelh, Lucinos~enwiki, GeeJo, Długosz, Nikkimaria, RodVance, Groyolo, MacsBug, SmackBot, Melchoir, Jrockley, Winterheart, Tigerhawkvok, Colonies Chris, Scwlong, Audriusa, V1adis1av, QFT, LouScheffer, Wikiwikiwiki3~enwiki, Doug Bell, John, JorisvS, Groggy Dice, Simkiott, Happy-melon, CRGreathouse, Sahrin, Michael C Price, Headbomb, Stannered, WinBot, Maliz, DerHexer, Momojeng, Jotempe, Dzogchenpa, TXiKiBoT, Hqb, Andysoh, Someguy1221, Bentley4, Ptrslv72, Northfox, SieBot, Chandrahas9, BlueAzure, Martarius, Chieron, Arjayay, SkyLined, Addbot, Barak Sh, Legobot II, Robert Treat, AnomieBOT, Rubinbot, Materialscientist, Citation bot, W.stanovsky, Ender's Shadow Snr, Jan Krieg, GreenRoot, Citation bot 1, Tom.Reding, ZéroBot, Vbrun237, Timetraveler3.14, Brandmeister, Surajt88, Nerdok, Fbrugmans, Teaktl17, ClueBot NG, Widr, Helpful Pixie Bot, Bibcode Bot, RageOfGod, BG19bot, GregorDS, EnzaiBot, Matherforthewin, Osteologia, Cjean42, Monkbot, Aberlamps, Nidj123 and Anonymous: 76

- **Radioactive decay** *Source:* https://en.wikipedia.org/wiki/Radioactive_decay?oldid=682565072 *Contributors:* The Anome, Danny, Roadrunner, Mrwojo, Spiff~enwiki, Patrick, Ahoerstemeier, Andrewa, LittleDan, Kricke, Samw, Mxn, Smack, Hike395, HolIgor, Chuljin, Jitse Niesen, Audin, Furrykef, Populus, Omegatron, Topbanana, Pstudier, Finlay McWalter, PuzzletChung, Robbot, Romanm, Chancemill, Securiger, Merovingian, Pengo, Giftlite, Fudoreaper, Netoholic, Herbee, Everyking, Snowdog, Curps, Eequor, Jackol, Mmm~enwiki, Manuel Anastácio, Utcursch, Andycjp, LiDaoing, Antandrus, Beland, DragonflySixtyseven, Icairns, GeoGreg, Urhixidur, Syvanen, Olivier Debre, Deglr6328, Kate, Running, Mike Rosoft, Mormegil, Freakofnurture, Discospinster, Rydel, Rama, Vsmith, Mjpieters, Mani1, Night Gyr, Bender235, ESkog, Sunborn, Tompw, El C, J-Star, Lankiveil, Joanjoc~enwiki, Hayabusa future, RoyBoy, Orestes~enwiki, Grick, Bobo192, Stesmo, Smalljim, Indio~enwiki, Cohesion, Kjkolb, Nsaa, Storm Rider, Alansohn, Mr Adequate, AjAldous, Seans Potato Business, Ynhockey, Velella, Harej, RainbowOfLight, Dirac1933, Sciurinæ, Mikeo, DV8 2XL, Paraphelion, Zntrip, Ocollard, StradivariusTV, Duncan.france, Miss Madeline, CharlesC, Wdanwatts, Jacj, Qwertyus, Jclemens, Scuzzman, Martinevos~enwiki, Rjwilmsi, Jmcc150, Nneonneo, Bubba73, Watcharakorn, Lionelbrits, Ground Zero, Old Moonraker, RexNL, Kolbasz, Dalef, Fresheneesz, Guliolopez, Gwernol, Roboto de Ajvol, Wavelength, Phmer, Kymacpherson, RussBot, Jengelh, Shawn81, Kerowren, David Woodward, Gaius Cornelius, CambridgeBayWeather, Rsrikanth05, Bovineone, Tungsten, Grafen, Jaxl, Welsh, ONEder Boy, Ino5hiro, DJ John, Lomn, Scottfisher, DeadEyeArrow, Jeremy Visser, Ignitus, Wknight94, FF2010, Light current, Sefarkas, Closedmouth, Јованвб, Reyk, CharlesHBennett, CWenger, Fourohfour, Caco de vidro,

Moomoomoo, Sbyrnes321, DVD R W, CIreland, Xtraeme, Eog1916, Itub, MacsBug, SmackBot, FocalPoint, Jclerman, Lcarsdata, Incnis Mrsi, KnowledgeOfSelf, Joonhon, Hydrogen Iodide, NoahWolfe, Jmulvey, Blue520, CMD Beaker, Jrockley, Yamaguchi⬜⬜, Gilliam, Carl.bunderson, TRosenbaum, Ati3414, Chris the speller, Bluebot, Kurykh, Agateller, Cadmium, MK8, Metacomet, Uthbrian, Reko, Sbharris, Rogermw, NYKevin, Can't sleep, clown will eat me, Ajaxkroon, Shalom Yechiel, Abyssal, V1adis1av, Ioscius, KaiserbBot, Rrburke, VMS Mosaic, Rsm99833, Addshore, Mrdempsey, Megamix, Flyguy649, Smooth O, Xyzzy n, Dreadstar, -Ozone-, Lcarscad, Cockneyite, Drphilharmonic, DMacks, Where, Bidabadi~enwiki, Cyberevil, Lambiam, SuperTycoon, Sanya, JoshuaZ, Accurizer, Minna Sora no Shita, IronGargoyle, 16@r, Ryulong, Peyre, Squirepants101, Dan Gluck, BranStark, Pegasus1138, CP\M, Freelance Intellectual, Fdp, Tawkerbot2, Chetvorno, Bstepp99, Conrad.Irwin, INkubusse, Xcentaur, RSido, Vyznev Xnebara, Nunquam Dormio, Solargenerator9.5, MarsRover, Leujohn, Smoove Z, Myasuda, J. Tyler, Island Dave, Quinnculver, Kanags, Gogo Dodo, HPaul, Mad-rick, Rracecarr, Skittleys, Christian75, FastLizard4, Gmoney650, The real avenger, Mikewax, Thijs!bot, Epbr123, Plmoknijb, Dougsim, Headbomb, Marek69, Deschreiber, Davidhorman, Meteoritekid, FourBlades, Stannered, Mentifisto, AntiVandalBot, Quintote, Jj137, Panu Petteri Höglund, Hanzoro5, Myanw, JAnDbot, Arch dude, Andonic, Xact, Snowynight, Acroterion, Geniac, Freedomlinux, Bongwarrior, VoABot II, AuburnPilot, Hillgentleman, JNW, Estonofunciona~enwiki, DMcanada, Klausok, Pixel ;-), Colinsweet, SparrowsWing, Indon, Animum, Dirac66, 28421u2232nfenfcenc, LorenzoB, Tswsl1989, JoergenB, Squidonius, Lewismatson, Chuckwatson, NatureA16, MartinBot, Mermaid from the Baltic Sea, Bus stop, R'n'B, Leyo, J.delanoy, Trusilver, Bogey97, Maurice Carbonaro, Cpiral, Gzkn, Stan J Klimas, DarkFalls, Dynetrekk~enwiki, Tarotcards, Pyrospirit, Sara0202, Chikinsawsage, Fountains of Bryn Mawr, Ohms law, Treisijs, Jim Swenson, Useight, Xiahou, RJASE1, Idioma-bot, ACSE, Cuzkatzimhut, Malik Shabazz, Deor, Matt1191, VolkovBot, ABF, VasilievVV, Philip Trueman, TXiKiBoT, Oshwah, Xenophrenic, Technopat, Hqb, Jcherbak, Someguy1221, Kirkpthompson, LeaveSleaves, Bearian, 0x539, Spiral5800, MichaelMorrill, Enigmaman, Yk Yk Yk, Bryan26, Synthebot, Falcon8765, Jluo, Sylent, Xxxlilbritxxx, Insanity Incarnate, Kehrbykid, Alytkin, Borne nocker, Brettdog, Deconstructhis, Starkrm, D. Recorder, Drawde22, SieBot, Tiddly Tom, Scarian, Viskonsas, Caltas, Soler97, Keilana, Nic92, TJHarrison, Oxymoron83, Faradayplank, Lightmouse, RW Marloe, Arnobarnard, Rj39pooch2, Nergaal, Babakathy, Martarius, ClueBot, HujiBot, Avenged Eightfold, GorillaWarfare, Fasette, Bobathon71, Pvineet131, The Thing That Should Not Be, Plastikspork, VsBot, Wysprgr2005, Denna Haldane, Skäpperöd, CounterVandalismBot, Akash1209, Dougdp, MindstormsKid, Jersey emt, Opaltehjerkzors, Robert Skyhawk, Jusdafax, Erebus Morgaine, Huzzy92, 06multan, Arjayay, Radiogenic, PhySusie, Iohannes Animosus, Francisco Albani, IXella007, Dekisugi, La Pianista, Thingg, Aitias, Jonverve, Plasmic Physics, Megachad, Party, OpusAtrum, Johnson-gray, MystBot, Angerfist~enwiki, Thatguyflint, Hobbema, CalumH93, Amezcackle, Addbot, Proofreader77, Chorro22, Magus732, Smb6009, Laurinavicius, CanadianLinuxUser, Leszek Jańczuk, WFPM, Cst17, LaaknorBot, PranksterTurtle, Exor674, Lordlosss2, Tide rolls, Jarble, Legobot, Luckas-bot, Yobot, TaBOT-zerem, Legobot II, Theropod, Amble, Ayrton Prost, Hurricaneguy, AnomieBOT, DemocraticLuntz, Killiondude, Jim1138, Piano non troppo, AdjustShift, Scuzzer, Law, Materialscientist, The High Fin Sperm Whale, Citation bot, E2eamon, Bob Burkhardt, LilHelpa, Xqbot, Transity, Capricorn42, Richarddgill, Webkinzgirl101, Omnipaedista, RibotBOT, Amaury, Doulos Christos, Eugene-elgato, Pumpmaster60, FrescoBot, Surv1v4l1st, Wusel007, LucienBOT, Wvilhellm, Tobby72, Pepper, Oldlaptop321, MagnaGraecia, Footyfanatic3000, HJ Mitchell, Cannolis, Citation bot 1, Arthree, Pinethicket, Edderso, 10metreh, Odyssey xg, A8UDI, Minivip, Meaghan, Double sharp, TobeBot, Trappist the monk, Lotje, Ndkartik, TheBFG, Mozi17, Comet Tuttle, Math.geek3.1415926, Dinamik-bot, Vrenator, Tobias1984, Bluefist, Specs112, SilverbladeGR, Cfsgfds, Fastilysock, Cutelyaware, Sampathsris, Minimac, TjBot, TomBeasley, KuanRyan, Androstachys, Alison22, DASHBot, TGCP, BotdeSki, John of Reading, WikitanvirBot, Lunaibis, RedHab, ScottyBerg, Yt95, RenamedUser01302013, Kulmeetster, Wikipelli, K6ka, Sydneyanders, JSquish, ZéroBot, John Cline, PBS-AWB, Mkevinjnr, Suslindisambiguator, Elio96, Gz33, QEDK, Aschwole, L Kensington, MonoAV, Maschen, Donner60, Scientific29, ChuispastonBot, RockMagnetist, Ryan Pianesi, Newtrend19, Petrb, ClueBot NG, Crazyman121, Littleal38, Verpies, Satellizer, Baseball Watcher, Slartibartfastibast, Widr, Dasetwundabal, Oddbodz, Helpful Pixie Bot, Ciro612, Strike Eagle, Calabe1992, Bibcode Bot, Jeraphine Gryphon, BG19bot, Teiu88, Northamerica1000, Wiki13, ElphiBot, MusikAnimal, Cynaide, Shampa1, Flying hippo705, Glevum, DynamicDino, Adebish, Zedshort, Hamish59, Mgoelzer, SfHuIcTk, Thegreatgrabber, Achowat, Imawesome12345678910, ArrakisFrance, 555snowy, Kisokj, Ezekiel25q, Wolf11235, Cyprien 1997, BrightStarSky, Apples122, Ultimatewikimaster12345, Reatlas, Joeinwiki, Cavisson, Tentinator, Awesome boss 69 69, Bond064, Jyotmankad, CloudStrifeNBHM, Jwratner1, Applezpi3, Genome0514, StevenD99, Bkilli1, Ilikethemchickenwing$, Andthewinneris...Cole, Zane7777, Shbew, Monkbot, UDDM, Vieque, Thenapster1426, TheFireRises, Micbattle064, Paul2lyfe, Amortias, Pacifist peeta, Radioactiveisreallyawesome, KasparBot, Cerberus123, Never gonna See me, Lexi sioz, Subhajit07, Soumik Pattanayak, Bigdaddyyyyy69 and Anonymous: 804

- **Electron capture** *Source:* https://en.wikipedia.org/wiki/Electron_capture?oldid=681029045 *Contributors:* Mav, Andre Engels, Imran, GaryW, Pstudier, Twang, Donarreiskoffer, Gentgeen, Romanm, SpellBott, Mikez, Art Carlson, Dratman, Icairns, B.d.mills, Hax0rw4ng, Newhoggy, Discospinster, Vsmith, Sunborn, Joanjoc~enwiki, Brim, Foobaz, Riana, DV8 2XL, Forteblast, Richard Arthur Norton (1958-), Benbest, Rjwilmsi, Chobot, DVdm, Tone, YurikBot, Spacepotato, Hairy Dude, Shawn81, Shaddack, Anomalocaris, Dna-webmaster, LeonardoRob0t, Incnis Mrsi, Jagged 85, Betacommand, Sbharris, V1adis1av, BIL, Drphilharmonic, Daniel.Cardenas, Untitleduser, C.jeynes, Diverman, Magere Hein, Icek~enwiki, Michael C Price, Headbomb, Hcobb, Roches, Dirac66, LorenzoB, Vinograd19, AstroHurricane001, Howa0082, Yonidebot, Jutiphan, Vatic7, Sheliak, VolkovBot, TXiKiBoT, Pamputt, SieBot, YonaBot, Flyer22, ClueBot, Cmj91uk, SchreiberBike, Oldnoah, NellieBly, SkyLined, Debzer, Addbot, LaaknorBot, Zorrobot, Skippy le Grand Gourou, Luckas-bot, AnomieBOT, ArthurBot, Xqbot, GrouchoBot, FrescoBot, PigFlu Oink, Minivip, Miracle Pen, AndyHe829, MartinThoma, Bibcode Bot, Snow Rise, Eio, Zedshort, Paní Slepičková, JPBrod, Maysens, BsGTeo, Spyglasses, Meteor sandwich yum, Monkbot, Haveasweater, Jsaur, Wqwt, Alma.f.r, KasparBot and Anonymous: 53

- **Quantum chromodynamics** *Source:* https://en.wikipedia.org/wiki/Quantum_chromodynamics?oldid=678061083 *Contributors:* AxelBoldt, CYD, Zundark, Youandme, Ewen, Stevertigo, Michael Hardy, Ahoerstemeier, Whkoh, Emperorbma, Jitse Niesen, Phys, Robbot, Fredrik, Ojigiri~enwiki, Seth Ilys, Alan Liefting, Giftlite, JamesMLane, Monedula, Xerxes314, JeffBobFrank, Jason Quinn, Elroch, Icairns, Sam Hocevar, Lumidek, Sctfn, Eep², David Schaich, JonL, Goplat, AdamSolomon, Pt, El C, CDN99, Robotje, Slicky, Physicistjedi, Azn king28, Fwb22, Guy Harris, Ricky81682, TenOfAllTrades, Skyring, Kusma, Alai, Mpatel, Betsythedevine, Mendaliv, VermillionBird, Rjwilmsi, Coemgenus, FlaBot, Thenewdeal87, Adoniscik, Algebraist, YurikBot, Wavelength, Bambaiah, Hairy Dude, Moto Perpetuo, Ohwilleke, JabberWok, Kirill Lokshin, Spike Wilbury, BlackAndy, Thiseye, CecilWard, Voidxor, Zzuuzz, Banus, Finell, SmackBot, Henriok, Vald, ProveIt, GaeusOctavius, Chris the speller, Bluebot, TimBentley, Complexica, Colonies Chris, Modest Genius, Berland, Grover cleveland, Garry Denke, TriTertButoxy, DJIndica, Jaganath, RoboDick~enwiki, NNemec, Slakr, Ryulong, Tawkerbot2, Memetics, Capefeather, Runningonbrains, Cydebot, DavidMcCabe, Headbomb, WVhybrid, Noclevername, Escarbot, Salgueiro~enwiki, Shambolic Entity, Andonic, Hut 8.5, Pkoppenb, .anacondabot, Robomojo, Corvidaecorvus, Maliz, Connor Behan, TechnoFaye, R'n'B, HEL, DrKiernan, Acalamari, Shomroni, Lseixas, Skullfunk,

GrahamHardy, Idioma-bot, Sheliak, Cuzkatzimhut, VolkovBot, TXiKiBoT, Calwiki, Rei-bot, Saibod, KP-Adhikari, Ptrslv72, SieBot, Dawn Bard, Likebox, Anchor Link Bot, ClueBot, WDavis1911, Pechmerle, PixelBot, Brews ohare, Chrisarnesen, XLinkBot, SilvonenBot, SkyLined, Truthnlove, Addbot, DOI bot, AnnaFrance, SpBot, Lightbot, Zorrobot, Legobot, Luckas-bot, Yobot, Tamtamar, Nallimbot, Citation bot, Lil-Helpa, Info21, Chrisfox8, Pra1998, Petros000, FrescoBot, Ecuqkindler, Timmeken, Ganondolf, Meier99, Heurisko, Earthandmoon, Tarsilia, McSaks, Autumnalmonk, EmausBot, Mnkyman, Wikipelli, Brazmyth, Quondum, Aschwole, Rcsprinter123, Maschen, Fwilczek, RolteVolte, Neduard, QuantumSquirrel, Teaktl17, ClueBot NG, Helpful Pixie Bot, Bibcode Bot, Dalit Llama, BG19bot, PhnomPencil, Vkpd11, Snow Blizzard, Cjean42, Joeinwiki, Trompedo, KasparBot and Anonymous: 137

- **Proton therapy** *Source:* https://en.wikipedia.org/wiki/Proton_therapy?oldid=679584928 *Contributors:* Kku, Julesd, David.Monniaux, HaeB, Xanzzibar, DocWatson42, Dratman, Curps, Erich gasboy, Beland, Maikel, Rich Farmbrough, Mashford, Pmetzger, Afed, Bobo192, Brim, Arcadian, Drw25, Jonathunder, Jhertel, Dental, BRW, Wikicaz, TenOfAllTrades, Zereshk, Pol098, Tabletop, SCEhardt, Mandarax, Graham87, Rjwilmsi, Corto, Gurch, Intgr, Tdvorak~enwiki, RussBot, JabberWok, Asacarny, Bbbozzz, Ageekgal, Esprit15d, Meegs, SmackBot, Benjaminevans82, Miquonranger03, Rpspeck, Sbharris, Alphathon, Reuvenmlerner, Pwjb, Tereufuk, Erwin, Dicklyon, TwistOfCain, Sopila2, Daedalus969, Mauricev, Cydebot, Rifleman 82, HPaul, Hopping, Nismo334, CopperKettle, J.Ring, Ksimmons8888, Porqin, Allahyar, Lenarzt, Sonicko, MER-C, Hydro, Haricotvert, Ph.eyes, Sortofscientist, Drcoop, Nbauman, Rod57, KylieTastic, Getoutandrun, Jamesontai, Ottershrew, Tinstaafl, Philip Trueman, Jtexada, Michaeldsuarez, Aci20, Doc James, AlleborgoBot, Gknor, Neutralhomer, Alexbrn, JsePrometheus, Bgordski, Optivusprime, Denisarona, Kjohnson47, LGBOUCHET, Cnsjones, SeeACure, PixelBot, Jamiecorbett, Vanished User 1004, DumZiBoT, RexxS, XLinkBot, Holoeconomics, Sdfsdfsdfsdfdsdfsdf, MystBot, Addbot, Fieldday-sunday, MarkFilipak, Proxima Centauri, Dkornguth, Bonewith, Jesssoul, Yobot, Amirobot, Jjsu, DiverDave, AnomieBOT, Jim1138, InsufficientData, Piano non troppo, Yachtsman1, Citation bot, Xqbot, Moneytoo, Cortamears, McKeesRocks, FrescoBot, LucienBOT, MC6508, Mdphd2012, Jonesey95, Wally2121, Trappist the monk, Yunshui, AE1978, VmZH88AZQnCjhT40, Difu Wu, Askedradio8, RjwilmsiBot, Wantdouble, KuduIO, Brandmeister, ClueBot NG, Selbytec, Agreda1, FRSC Chemist, Labpluto123, Cphreak, Exurbis67, Helpful Pixie Bot, Sylvain.coppens, BG19bot, JohnChrysostom, BattyBot, Mogism, Cerabot~enwiki, UseTheCommandLine, Merxistan, PE CSIntell, Ruby Murray, Jodosma, P2rhodes, Jdmm72, Safsaftunis, Helper1976, Zunpre, Yeon-Joo Kim, Suz1806, K9re11, Mf.partners, Monkbot, Vassili.Petrovitch-J, Lime0life, Wiki CRUK John, Nina Bardi de Alvarez, IChevako, Glory789, Spiderjerky, Crabtoast1, Zabineph, 2Kaepsele56, Yeahsangs5 and Anonymous: 171

- **Hydron (chemistry)** *Source:* https://en.wikipedia.org/wiki/Hydron_(chemistry)?oldid=667303209 *Contributors:* Stone, Graeme Bartlett, Julianonions, DePiep, Okedem, Itub, Incnis Mrsi, ZerodEgo, Chris the speller, Sbharris, A876, Rifleman 82, Christian75, Washod, Magioladitis, Dirac66, Leyo, Broadbot, BotKung, W4chris, SieBot, Puppy8800, Plasmic Physics, Addbot, DOI bot, Wickey-nl, LaaknorBot, Ehrenkater, Lightbot, Luckas-bot, Amirobot, Reindra, AnomieBOT, Chotu21, GrouchoBot, Firq, MastiBot, Mikespedia, December21st2012Freak, BogBot, Double sharp, 777sms, Armando-Martin, WikitanvirBot, Dcirovic, JSquish, ClueBot NG, BG19bot, Monkbot and Anonymous: 9

- **Proton nuclear magnetic resonance** *Source:* https://en.wikipedia.org/wiki/Proton_nuclear_magnetic_resonance?oldid=666909791 *Contributors:* Complex Analysis, Habrahamson, H Padleckas, Danny B-), Dtsang, Walkerma, Velella, Baltakatei, GregorB, V8rik, Rangek, Bgwhite, Kkmurray, Allens, Itub, SmackBot, Colonies Chris, Can't sleep, clown will eat me, Shalom Yechiel, Mothball, DMacks, Daniel.Cardenas, Molerat, G-W, Rifleman 82, Christian75, Magioladitis, ChemistHans, Quantockgoblin, Dirac66, CommonsDelinker, Nova5, AstroHurricane001, Nathansbenjamin, KingCuongL, PerryTachett, ImageRemovalBot, R. London Griffin, Dr.Soft, Addbot, Ettrig, Yobot, TaBOT-zerem, THEN WHO WAS PHONE?, Invest in knowledge, LucienBOT, Chutznik, Tom.Reding, Geoffrais, Aleckwyq, Superdoofer, Teuta9, BG19bot, MichaelCrocker, Spire music, Monkbot, Languagetooler, Carolineneil and Anonymous: 43

- **Proton–proton chain reaction** *Source:* https://en.wikipedia.org/wiki/Proton%E2%80%93proton_chain_reaction?oldid=676579786 *Contributors:* Carey Evans, Andre Engels, Roadrunner, Xavic69, Looxix~enwiki, Caid Raspa, Kimiko, Kbk, Rursus, Wereon, Lupo, Decumanus, Centrx, Harp, Foobar, Beland, Icairns, Sam Hocevar, Tsemii, ESkog, Vuo, Gene Nygaard, Ataru, Brownsteve, RichardWeiss, Uxh, Williamborg, Kevmitch, E2rd, BitterMan, Chobot, Witan, Hellbus, Thiseye, Lexicon, Reyk, Nekura, KnightRider~enwiki, SmackBot, Jrockley, Wykis, Sbharris, Bowlhover, A5b, Wikier.ko, Gobonobo, Fontenello, JorisvS, Uwe W., Newone, Happy-melon, Menswear, Petr Matas, CmdrObot, Syphondu, ProfessorPaul, A876, Islander, WISo, Michael C Price, Patrick O'Leary, Crum375, Epbr123, Headbomb, Escarbot, Orionus, Qwerty Binary, Antwan911, H3llbringer, Belg4mit, WolfmanSF, Geboy, Pagw, Glrx, Numbo3, Hans Dunkelberg, Sheliak, PNG crusade bot, Hqb, Broadbot, Newcomp, Briansacks, Hubbcapp, Anton Gutsunaev, Tjabell, Dynamitecow, Davidallred, Hyh1048576, Pplfichi, Debsuvra, Jotterbot, Wnt, SkyLined, Addbot, Njaelkies Lea, CanadianLinuxUser, Favonian, Lightbot, Luckas-bot, Yobot, AnomieBOT, USConsLib, ArthurBot, Xqbot, Abeshenkov, Mnmngb, Joaquin008, Dave3457, Ironboy11, D'ohBot, IVAN3MAN, DixonDBot, Xiaomao123, Dewritech, ZéroBot, Arbnos, Zitterbewejung, Brandmeister, Terraflorin, Whoop whoop pull up, ClueBot NG, Mark Zelinka, Bibcode Bot, AvocatoBot, Scientistmohamed, Samcstewart, Tony Mach, Monkbot, Leandervb, Isambard Kingdom, Hovereel, Dion10474 and Anonymous: 66

- **Proton spin crisis** *Source:* https://en.wikipedia.org/wiki/Proton_spin_crisis?oldid=666895257 *Contributors:* RJFJR, Christopher Thomas, Sperxios, Lambiam, הסרפד, Headbomb, AndyBloch, GoatGuy, Vegasprof, C933103, Yobot, SwisterTwister, AnomieBOT, Citation bot, ⁇⁇, Sae1962, Tom.Reding, Ofercomay, Timetraveler3.14, Brandmeister, Bibcode Bot, BG19bot, 786b6364, Harsh 2580, Tony Mach, Andyhowlett, Vk1987, Sashnik and Anonymous: 7

- **Antiproton** *Source:* https://en.wikipedia.org/wiki/Antiproton?oldid=676938934 *Contributors:* Mav, Hephaestos, Patrick, Jiang, Evercat, Saltine, AaronSw, Robbot, Millosh, Harp, Snowdog, Fleminra, Leonard G., Bodhitha, DragonflySixtyseven, KeithTyler, Mike Rosoft, Rich Farmbrough, Bender235, Goplat, Triona, Nk, Helix84, Cburnett, Saxifrage, Mindmatrix, Robert K S, Kralizec!, Kbdank71, Rjwilmsi, Strait, Mahlum~enwiki, Goudzovski, Wrightbus, Eric B, Whosasking, YurikBot, Bambaiah, Jimp, Evrik, Reyk, Mkossick, Attilios, SmackBot, FocalPoint, Hmains, Stevage, Sbharris, Rogermw, Pwjb, Andrei Stroe, Tom9729, Flip619, MenaceSan, Clarityfiend, Newone, Happy-melon, Thijs!bot, Headbomb, Northumbrian, JAnDbot, Z22, Khaosjr, Fellwalker57, HEL, Siryendor, C.Kalvin, Larryisgood, TXiKiBoT, Spinningspark, SieBot, Ghaller, ClueBot, Farras Octara, 2004williamsj, Carsrac, DumZiBoT, Ramisses, Vanished user k3rmwkdmn4tjna3d, SilvonenBot, SkyLined, Addbot, DOI bot, Lightbot, Legobot, Luckas-bot, Fraggle81, Citation bot, ArthurBot, Franco3450, Mnmngb, Paine Ellsworth, Steve Quinn, Ysyoon, ClickRick, Citation bot 1, TobeBot, Trappist the monk, 564dude, Earthandmoon, Ripchip Bot, WikitanvirBot, IncognitoErgoSum, Hhhippo, NicoWpa, Brandmeister, Whoop whoop pull up, ClueBot NG, Frietjes, Bibcode Bot, Slaughter182, Drizzt182, ChrisGualtieri, Pimpoosh, RobertLM78, Green Zero, Monkbot, Wikicology, KasparBot and Anonymous: 67

16.5.2 Images

- **File:1H_NMR_Ethyl_Acetate_Coupling_shown.png**_Source:_https://upload.wikimedia.org/wikipedia/commons/8/80/1H_NMR_Ethyl_ Acetate_Coupling_shown.png_License:_CC BY-SA 3.0_Contributors:_This file was derived from1H NMR Ethyl Acetate Coupling shown - 2.png: <ahref='//commons.wikimedia.org/wiki/File:1H_NMR_Ethyl_Acetate_Coupling_shown_-_2.png' class='image'><img alt='1H NMR Ethyl Ac -etate Coupling shown - 2.png' src='https://upload.wikimedia.org/wikipedia/commons/thumb/1/19/1H_NMR_Ethyl_Acetate_Coupling_shown _-_2.png/50px-1H_NMR_Ethyl_Acetate_Coupling_shown_-_2.png' width='50' height='36' srcset='https://upload.wikimedia.org/wikipedia/ commons/thumb/1/19/1H_NMR_Ethyl_Acetate_Coupling_shown_-_2.png/75px-1H_NMR_Ethyl_Acetate_Coupling_shown_-_2.png 1.5x, https://upload.wikimedia.org/wikipedia/commons/thumb/1/19/1H_NMR_Ethyl_Acetate_Coupling_shown_-_2.png/100px-1H_NMR_Ethyl_ Acetate_Coupling_shown_-_2.png 2x' data-file-width='631' data-file-height='456' /></a>_Original artist:_1H_NMR_Ethyl_Acetate_Coupling_ shown.GIF:T.vanschaik

- **File:Alfa_beta_gamma_radiation.svg** *Source:* https://upload.wikimedia.org/wikipedia/commons/d/d6/Alfa_beta_gamma_radiation.svg *License:* CC BY 2.5 *Contributors:* Traced from this PNG image. *Original artist:* User:Stannered

- **File:Alpha_Decay.svg** *Source:* https://upload.wikimedia.org/wikipedia/commons/7/79/Alpha_Decay.svg *License:* Public domain *Contributors:* This vector image was created with Inkscape. *Original artist:* Inductiveload

- **File:Ambox_important.svg** *Source:* https://upload.wikimedia.org/wikipedia/commons/b/b4/Ambox_important.svg *License:* Public domain *Contributors:* Own work, based off of Image:Ambox scales.svg *Original artist:* Dsmurat (talk · contribs)

- **File:Antiproton_accumulator_at_Femilab.jpg** *Source:* https://upload.wikimedia.org/wikipedia/commons/6/6d/Antiproton_accumulator_ at_Femilab.jpg *License:* CC BY-SA 4.0 *Contributors:* Own work *Original artist:* Z22

- **File:Atomic_rearrangement_following_an_electron_capture.svg** *Source:* https://upload.wikimedia.org/wikipedia/commons/b/b1/Atomic_ rearrangement_following_an_electron_capture.svg *License:* CC BY-SA 4.0 *Contributors:* Own work *Original artist:* Pamputt

- **File:Baryon_decuplet.svg** *Source:* https://upload.wikimedia.org/wikipedia/commons/f/f6/Baryon_decuplet.svg *License:* Public domain *Contributors:* Own work (Original text: *self-made*) *Original artist:* Wierdw123 at English Wikipedia

- **File:CERN_LHC_Tunnel1.jpg** *Source:* https://upload.wikimedia.org/wikipedia/commons/f/fc/CERN_LHC_Tunnel1.jpg *License:* CC BY- SA 3.0 *Contributors:* Own work *Original artist:* Julian Herzog (website)

- **File:Commons-logo.svg** *Source:* https://upload.wikimedia.org/wikipedia/en/4/4a/Commons-logo.svg *License:* ? *Contributors:* ? *Original artist:* ?

- **File:Comparison_of_dose_distributions_between_IMPT_(right)_and_IMRT_(left).jpg** *Source:* https://upload.wikimedia.org/wikipedia/ commons/e/e3/Comparison_of_dose_distributions_between_IMPT_%28right%29_and_IMRT_%28left%29.jpg *License:* CC BY 2.0 *Contributors:* http://www.ro-journal.com/content/3/1/4/figure/F2?highres=y *Original artist:* Taheri-Kadkhoda et al. Radiation Oncology 2008 3:4 doi:10.1186/1748-717X-3-4

- **File:Comparison_of_dose_profiles_for_proton_v._x-ray_radiotherapy.png** *Source:* https://upload.wikimedia.org/wikipedia/commons/ 3/37/Comparison_of_dose_profiles_for_proton_v._x-ray_radiotherapy.png *License:* CC BY-SA 3.0 *Contributors:* Graph created by me using CorelDraw X3.
  Previously published: (none) *Original artist:* MarkFilipak

- **File:Crookes_tube_xray_experiment.jpg** *Source:* https://upload.wikimedia.org/wikipedia/commons/1/10/Crookes_tube_xray_experiment. jpg *License:* Public domain *Contributors:* Downloaded 2007-12-23 from <a data-x-rel='nofollow' class='external text' href='http://books.google. com/books?id=whc4AAAAMAAJ,,&,,pg=PT5'>William J. Morton and Edwin W. Hammer (1896) *The X-ray, or Photography of the Invisible and its value in Surgery*, American Technical Book Co., New York, fig. 54</a> on Google Books *Original artist:* William J. Morton

- **File:DecayRate_vs_Solar_Time.png** *Source:* https://upload.wikimedia.org/wikipedia/commons/d/d3/DecayRate_vs_Solar_Time.png *License:* Public domain *Contributors:* ? *Original artist:* ?

- **File:Elementary_particle_interactions_in_the_Standard_Model.png**_Source:_https://upload.wikimedia.org/wikipedia/commons/a/a7/_ particle_interactions_in_the_Standard_Model.png *License:* CC0 *Contributors:* Own work *Original artist:* Eric Drexler

- **File:Flag_of_Canada.svg** *Source:* https://upload.wikimedia.org/wikipedia/en/c/cf/Flag_of_Canada.svg *License:* PD *Contributors:* ? *Original artist:* ?

- **File:Flag_of_France.svg** *Source:* https://upload.wikimedia.org/wikipedia/en/c/c3/Flag_of_France.svg *License:* PD *Contributors:* ? *Original artist:* ?

- **File:Flag_of_Germany.svg** *Source:* https://upload.wikimedia.org/wikipedia/en/b/ba/Flag_of_Germany.svg *License:* PD *Contributors:* ? *Original artist:* ?

- **File:Flag_of_Italy.svg** *Source:* https://upload.wikimedia.org/wikipedia/en/0/03/Flag_of_Italy.svg *License:* PD *Contributors:* ? *Original artist:* ?

- **File:Flag_of_Japan.svg** *Source:* https://upload.wikimedia.org/wikipedia/en/9/9e/Flag_of_Japan.svg *License:* PD *Contributors:* ? *Original artist:* ?

- **File:Flag_of_Korea_1882.svg** *Source:* https://upload.wikimedia.org/wikipedia/commons/c/c4/Flag_of_Korea_%281882-1910%29.svg *License:* Public domain *Contributors:* http://bugs.freeciv.org/ *Original artist:* Daniil Ivanov

- **File:Flag_of_Poland.svg** *Source:* https://upload.wikimedia.org/wikipedia/en/1/12/Flag_of_Poland.svg *License:* Public domain *Contributors:* ? *Original artist:* ?

- **File:Periodic_Table_Stability_&_Radioactivity.png** *Source:* https://upload.wikimedia.org/wikipedia/commons/c/c4/Periodic_Table_Stab %26_Radioactivity.png*License:*CC BY-SA2.5*Contributors:*https://commons.wikimedia.org/wiki/File:c _Table_Radioactivity.svg*Original artist:*Alessio Rolleri(et al),Lexi sioz

- **File:Pierre_and_Marie_Curie.jpg** *Source:* https://upload.wikimedia.org/wikipedia/commons/6/6c/Pierre_and_Marie_Curie.jpg *License:* Public domain *Contributors:* hp.ujf.cas.cz (uploader=--Kuebi 18:28, 10 April 2007 (UTC)) *Original artist:* Unknown

- **File:Portal-puzzle.svg** *Source:* https://upload.wikimedia.org/wikipedia/en/f/fd/Portal-puzzle.svg *License:* Public domain *Contributors:* ? *Original artist:* ?

- **File:PositronDiscovery.jpg** *Source:* https://upload.wikimedia.org/wikipedia/commons/6/69/PositronDiscovery.jpg *License:* Public domain *Contributors:* Anderson, Carl D. (1933). "The Positive Electron". *Physical Review* **43** (6): 491–494. DOI:10.1103/PhysRev.43.491. *Original artist:* Carl D. Anderson (1905–1991)

- **File:Predicted_proton_NMR_of_1,4-dimethylbenzene_from_ChemDraw._The_ratio_of_signal_strengths_of_proton_A_and_ proton_B_equals_to_their_molar_ratio_in_the_molecule..png**Source:https://upload.wikimedia.org/wikipedia/commons/c/c2/Predicted_ proton_NMR_of_1%2C4-dimethylbenzene_from_ChemDraw._The_ratio_of_signal_strengths_of_proton_A_and_proton_B_equals_to_their_ molar_ratio_in_the_molecule..png*License:*CC BY-SA 4.0*Contributors:*Own work*Original artist:*Carolineneil

- **File:Proton-Proton_III_chain_reaction.svg**Source:https://upload.wikimedia.org/wikipedia/commons/c/c4/Proton-Proton_III_chain_ reaction.svg*License:*CC BY-SA 3.0*Contributors:*http://en.wikipedia.org/wiki/File:Proton-Proton_II_chain_reaction.svghttps://en.wikipedia.org/ wiki/File:Proton-Proton_III_chain_reaction.png*Original artist:*Hovereel

- **File:Proton-Proton_II_chain_reaction.svg** *Source:* https://upload.wikimedia.org/wikipedia/commons/1/17/Proton-Proton_II_chain_reaction. svg *License:* CC BY-SA 3.0 *Contributors:*

- FusionintheSun.svg *Original artist:* FusionintheSun.svg: Borb

- **File:Proton_decay.svg** *Source:* https://upload.wikimedia.org/wikipedia/commons/8/85/Proton_decay.svg *License:* CC BY-SA 3.0 *Contributors:* Own work *Original artist:* Cjean42

- **File:Proton_proton_cycle.svg** *Source:* https://upload.wikimedia.org/wikipedia/commons/a/ac/Proton_proton_cycle.svg *License:* CC BY 2.5 *Contributors:* file:Proton proton cycle.png *Original artist:* Dorottya Szam

- **File:QCD.svg** *Source:* https://upload.wikimedia.org/wikipedia/commons/2/2b/QCD.svg *License:* CC BY-SA 3.0 *Contributors:* Own work *Original artist:* Cjean42

- **File:Quark_structure_antiproton.svg** *Source:* https://upload.wikimedia.org/wikipedia/en/2/24/Quark_structure_antiproton.svg *License:* CC-BY-SA-3.0 *Contributors:*

  Self created using Inkscape from File:Quark_structure_proton.svg as a template. *Original artist:*

  <a href='//en.wikipedia.org/wiki/User:Spinningspark' title='User:Spinningspark'>**Spinning**Spark</a> real life identity: SHA-1 commitment ba62ca25da3fee2f8f36c101994f571c151abee7

- **File:Quark_structure_proton.svg** *Source:* https://upload.wikimedia.org/wikipedia/commons/9/92/Quark_structure_proton.svg *License:* CC BY-SA 2.5 *Contributors:* Own work *Original artist:* Arpad Horvath

- **File:Question_book-new.svg** *Source:* https://upload.wikimedia.org/wikipedia/en/9/99/Question_book-new.svg *License:* Cc-by-sa-3.0 *Contributors:*

  Created from scratch in Adobe Illustrator. Based on Image:Question book.png created by User:Equazcion *Original artist:*

  Tkgd2007

- **File:R-parity_violating_decay.svg** *Source:* https://upload.wikimedia.org/wikipedia/commons/7/71/R-parity_violating_decay.svg *License:* Public domain *Contributors:* en:Image:RPD.png *Original artist:* en:User:Maliz, User:Stannered

- **File:Radioactive.svg** *Source:* https://upload.wikimedia.org/wikipedia/commons/b/b5/Radioactive.svg *License:* Public domain *Contributors:* Created by Cary Bass using Adobe Illustrator on January 19, 2006. *Original artist:* Cary Bass

- **File:Radioactive_decay_modes.svg** *Source:* https://upload.wikimedia.org/wikipedia/commons/7/71/Radioactive_decay_modes.svg *License:* GFDL *Contributors:* Own work *Original artist:* MarsRover

- **File:Radioactivity_and_radiation.png** *Source:* https://upload.wikimedia.org/wikipedia/commons/6/6e/Radioactivity_and_radiation.png *License:* CC BY-SA 3.0 *Contributors:* Own work *Original artist:* Doug Sim

- **File:Rutherford_1911_Solvay.jpg** *Source:* https://upload.wikimedia.org/wikipedia/commons/3/3b/Rutherford_1911_Solvay.jpg *License:* Public domain *Contributors:* ? *Original artist:* ?

- **File:Standard_Model_Feynman_Diagram_Vertices.png** *Source:* https://upload.wikimedia.org/wikipedia/commons/7/75/Standard_Model_ Feynman_Diagram_Vertices.png *License:* CC BY-SA 3.0 *Contributors:* I made it in Adobe Illustrator *Original artist:* Garyzx

- **File:Standard_Model_of_Elementary_Particles.svg**Source:* https://upload.wikimedia.org/wikipedia/commons/0/00/Standard_Model_of_ Elementary_Particles.svg *License:* CC BY 3.0 *Contributors:* Own work by uploader, PBS NOVA [1], Fermilab, Office of Science, United States Department of Energy, Particle Data Group *Original artist:* MissMJ

- **File:Stylised_Lithium_Atom.svg** *Source:* https://upload.wikimedia.org/wikipedia/commons/e/e1/Stylised_Lithium_Atom.svg *License:* CC-BY-SA-3.0 *Contributors:* based off of Image:Stylised Lithium Atom.png by Halfdan. *Original artist:* SVG by Indolences. Recoloring and ironing out some glitches done by Rainer Klute.

- **File:Table_isotopes_en.svg** *Source:* https://upload.wikimedia.org/wikipedia/commons/c/c4/Table_isotopes_en.svg *License:* CC BY-SA 3.0 *Contributors:*

- Table_isotopes.svg *Original artist:* Table_isotopes.svg: Napy1kenobi

- **File:The_incomplete_circle_of_everything.svg** *Source:* https://upload.wikimedia.org/wikipedia/commons/0/0d/The_incomplete_circle_of_everything.svg *License:* CC BY 3.0 *Contributors:* Own work *Original artist:* Zhitelew

- **File:Wikibooks-logo-en-noslogan.svg** *Source:* https://upload.wikimedia.org/wikipedia/commons/d/df/Wikibooks-logo-en-noslogan.svg *License:* CC BY-SA 3.0 *Contributors:* Own work *Original artist:* User:Bastique, User:Ramac et al.

- **File:Wiktionary-logo-en.svg** *Source:* https://upload.wikimedia.org/wikipedia/commons/f/f8/Wiktionary-logo-en.svg *License:* Public domain *Contributors:* Vector version of Image:Wiktionary-logo-en.png. *Original artist:* Vectorized by Fvasconcellos (talk · contribs), based on original logo tossed together by Brion Vibber

- **File:Yes_check.svg** *Source:* https://upload.wikimedia.org/wikipedia/en/f/fb/Yes_check.svg *License:* PD *Contributors:* ? *Original artist:* ?

16.5.3 Content license

- Creative Commons Attribution-Share Alike 3.0